WORKSHOP CHATTER

Duke of Edinburgh Trophy winner Miss Cherry Hinds with two of her models, a 1905 Fire Engine which took 3½ years time to build and, below her Allchin Traction Engine which occupied seven winters.

Workshop Chatter

A Bedside Book
for
Model Engineers

by
Martin Evans

Model & Allied Publications
Argus Books, 14 St. James Road
Watford, Herts.

First Published 1979

ISBN 0 85242 608 9

ALSO BY MARTIN EVANS

Manual of Model Steam Locomotive Construction
Outdoor Model Railways
Model Locomotive Boilers
Model Locomotive Valve Gears
Rob Roy
Simplex
Caribou

Railway History

Atlantic Era
Pacific Steam
Inverness to Crewe

Typeset by Inforum Ltd., Portsmouth
Printed and bound in Great Britain
by A. Wheaton & Co. Ltd., Exeter

Contents

Introduction

WHEN on visits to exhibitions of model engineering, model locomotive rallies and the like, I am often approached by people who have never known the joys of our unique hobby with the question 'What use is it?'

This little book has been written primarily to answer that question, for I believe that if the model engineering hobby was better understood, thousands more would try their hand at model-making, perhaps join a model engineering society and enjoy the many amenities that these organizations can offer.

Model engineering is a hobby for young and old, and both sexes too; for is not one of the finest model engineers of the present time a lady, who has carried off the Duke of Edinburgh Trophy at the Model Engineer Exhibition in open competition with the best male model builders?

It is true that practising model engineers are not often, nowadays, accused of 'playing trains', nevertheless the suspicion may still remain in some quarters that we are wasting our time (and money!). I hope that this short work may dispel such thoughts.

I have called this little book 'Workshop Chatter', and I hope that it will explain to the layman what makes the model engineer 'tick'. I hope also that it will make pleasant reading for all those interested in our fascinating hobby; it may, who knows, even bring a smile or two to the dedicated model engineer, or to the reader who, for one reason or another, is unable at present to follow his favourite hobby.

Martin Evans — Eydon, 1979

A popular 3½ in. gauge locomotive. "Rob Roy" — one of the author's designs. Quite easy to build, but with a good performance on the track. The model was based on a Caledonian Railway dockyard shunter. This example was built by Bertie Green of Harting, as a first attempt.

On the model engineering hobby

IT WOULD be interesting to know when the hobby of model engineering first started. Models of a simple kind were of course built many centuries ago; but I think it would be true to say that the hobby as an organized affair really started in 1898, when that great man Percival Marshall founded the magazine *Model Engineer*, and together with some enthusiastic friends, formed the first model engineering club in the world, the Society of Model and Experimental Engineers, as it is now known.

Since those early days, clubs and societies have sprung up in most of our large cities, and even in some quite small towns. Societies have also been formed in many countries overseas, notably in the U.S.A., Canada, South Africa, Australia, New Zealand, France, Belgium, Germany, Russia, and so on.

But the reader must not run away with the idea that model engineering is only carried out by members of organized societies. Far from it; there are many more 'lone hands' than there are members of clubs, though we do not, for obvious reasons, hear so much about their activities, nor do we often have the opportunity of seeing and admiring their models. But of that, more anon.

At one time, and not so long ago, we who build models were accused of making toys. When we operated our model steam locomotives, we were told that we were 'playing trains', and that we were probably in our second childhood. Fortunately, people know better nowadays, thanks to public exhibitions of the work of model engineers, to radio and television programmes, and to the outdoor activities of the clubs themselves, with their rallies and track days. Today, the non-technically minded are at last beginning to understand what the hobby is all about.

It is often asked why model engineers tend to specialize in one particular type of model, though there are of course some exceptions to this. The model steam locomotive is almost certainly the most popular of all models at the present time, and will probably continue to be so for

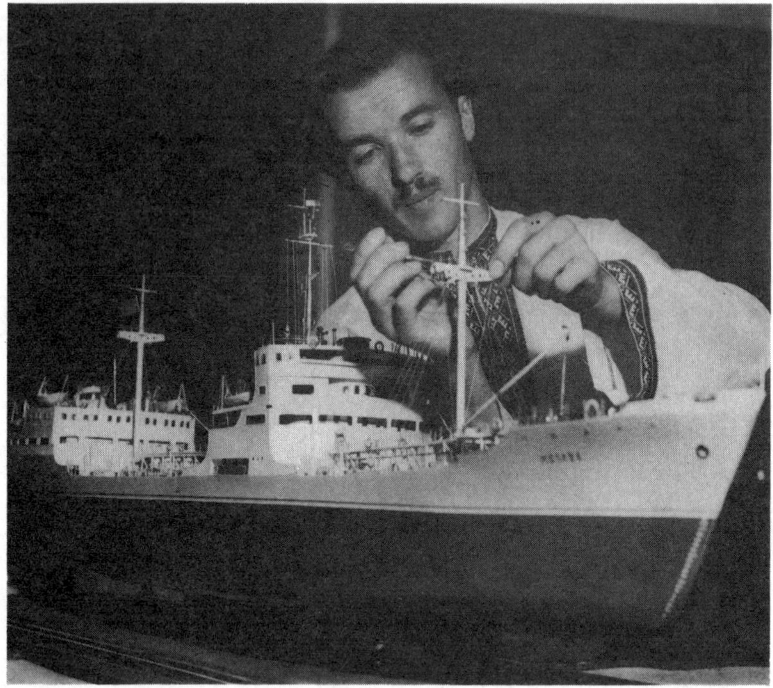

Alexander Vicherov from the Ukraine with his 1/100 scale model of a Russian tanker "Mockba", the prototype of which is in service in the Black Sea.

many years yet, in spite of the demise of the steam locomotive as far as full-size railways are concerned. After locomotives, I think model ships and power boats are next in popularity, followed by traction-engines, steam-rollers and similar machines. My own theory about this 'specialization' is that we tend to build models of the full-size machine that we loved as boys. This is certainly true in my own case.

I spent my early years in a large, old-fashioned house overlooking the old Great Eastern Railway's main line to Cambridge. From my bedroom window, I could see the four-track main line, and that remarkable arrangement of tracks known as Coppermill Junction, where the direct line from Liverpool Street swept down from a bridge over the River Lea, round a sharp curve to join the lines from the big Temple Mills marshalling yard — the tracks to Chingford continuing over the main lines. This was an ideal place to watch train movements, as in addition to the very heavy suburban traffic, there were plenty of expresses and 'semi-fasts', and in those days, a very heavy goods traffic, including the occasional 'foreigner' in the shape of a Midland or LMS freight train.

Perhaps my very earliest recollection is of persuading my grandmother to take me — then about six years old — down to the line, which involved crossing the River Lea twice, and running the gauntlet of the horse-drawn timber barges with their long ropes on either side of the water; there were no diesel barges then, all the barges were drawn by huge shire horses.

Having left the river and its tributaries behind, we would then have to cross an expanse of rough grass-land, usually boggy in winter; this formed part of the so-called Walthamstow Marshes. Finally we would come to the main lines, which crossed Coppermill Stream on two steel girder bridges, which were built so low over the pathway that headroom could not have been more than four feet. It is just the same today.

Some years later, I would set off for the railway alone, on every possible occasion, armed with notebooks and pencils, and copious notes would be made of every locomotive passing. My particular favourites were the '1500' class 4-6-0s, which I always thought looked particularly fine in their (then) brand new LNER applegreen livery, with the spokes of their driving wheels flashing round past the slots in the valances. I also admired the many classes of 0-6-0 goods engines, the big Great Eastern J 19s and J 20s, and later on, the Gresley J 39s, which I thought extremely handsome and powerful-looking.

After some years, I acquired a cycle, and was able to travel further afield in my search for locomotives; first there would be visits to Temple Mills Sidings, then westwards to the Great Northern main line, where

Gallopers, or roundabouts make most attractive models and here 9-year old Nicola Puttock looks at the entry of W.H. Heather at the Model Engineer Exhibition.

the famous Great Northern 'Atlantics' could be seen hurtling through Wood Green and Harringay, and where Gresley's first 'Pacifics' were creating quite a stir.

But how did I change from a mere 'number-taker' to a builder of model locomotives? It happened this way. I was passing the local newsagent's shop — I must have been about thirteen at the time — when I spotted in his window a magazine which dealt exclusively with model railways. This was of course the now well-known *Model Railway News*. Fearful lest that copy in the window should be the last, or perhaps already spoken for, I dashed into the shop, and was very soon devouring (I can't think of a more appropriate word!) Leyland-Barrett on Great Western locomotives, 'Sparks' on wagon-building for Gauge 0, Sir Eric Hutchison on LNER special wagons, and much else, all completely new to me.

It was not long before I had found my way to the premises of one of the bigger advertisers in the *MRN* — Messrs Bonds o' Euston Road. Bonds' old shop in the Euston Road was a veritable model maker's Mecca. The old premises I recall were narrow, but very deep, and the

A working model tappet loom built by Mr. T.W. Millward of Manchester.

A model tricycle made by the late A.F. Weaver.

Tramcars make fine models, and this L.C.C. class E.1. car was built by the late F.J. Roche.

A model of the Monosoupape rotary engine by J. Loudon, winner of a Silver Medal and the Bradbury-Winter Memorial Trophy at the Model Engineer Exhibition.

assistants seemed to me to know all about model railways and model engineering.

Like so many model railway enthusiasts, my first attempts were simple wagons for Gauge 0 — open wagons with wooden bodies and metal parts which I was able to buy ready-made — wheels, axleguards and the like. The idea of building a locomotive seemed then impossibly remote. I remember thinking that I could never learn to solder, so how could I possibly build a scale locomotive? I would have to save up and have one built by one on the specialist firms advertising in *MRN*.

About two years later, I became the proud possessor of my first scale locomotive, an LNER N 2 class 0-6-2 tank engine, hand-made in tinplate, and fitted with Leeds Model Co. mechanism — 12 volt too: I was not going to be fobbed off with a mere 6 volter! It was as a result of acquiring this locomotive, and having at the time nowhere to run it, that I plucked up my courage and applied to join the (London) Model

Railway Club, which at that time had its headquarters at a school in the Tottenham Court Road. And it was while attending the meetings of this club that I heard about the Model Engineer Exhibition.

I shall have more to say about the Model Engineer Exhibition in a later chapter. Suffice it to say now that my first visit to this exhibition really opened my eyes as far as the hobby was concerned. I remember that one of my first reactions was to realize the much greater challenge that the 'live steam' model locomotive presented, when compared with the electrically-driven 'steam-outline' model. But for various reasons, I did not give up my Gauge 0 models immediately; it was some three years later, just before the outbreak of the Second World War, that I made my first venture into model engineering proper.

To anyone used to Gauge 0 or 00 railway models, even 2½ in. gauge steam-driven models seem frighteningly large, and much though I admired the 5 in. gauge engines that I saw in steam on the SMEE track, I thought that from every point of view, they were far beyond me, and my first effort in 'live steam' proved to be LBSC's *Mary Ann*, a 2½ in. gauge version of the LNER J 39 class 0-6-0, one of those engines I had admired so much in earlier years. Soon after this I attempted a 3½ in. gauge L.N.E.R. *Green Arrow*, which, due to the war, never progressed beyond the chassis stage.

Those who take up model engineering today may do so for very different reasons. For one thing, the hobby itself is much better known

A model of a quick-firing naval gun by W.E. Radford.

An unfinished model of the BRM racing engine under construction by Professor D.H. Chaddock of Quorn, Leics.

than it was, thanks more than anything to the activities of the clubs. The newcomer may be fascinated by some model seen at one of the many exhibitions organized by model engineering societies, and decide to 'have a go' himself; he may have visited one of the many rallies organized by the National Traction Engine Club and similar bodies, and acquired a strong desire to build a model of one of the traction-engines or steam wagons which turn out on these occasions. Then there are those who have no great interest in models of any kind, but are fascinated by machines or tools — generally the small centre lathe — and whose hobby is the perfection of their machines and the designing and building of equipment and attachments for them.

At this point, perhaps we should ask ourselves exactly which branches of model-making or kindred activities are really embraced by the term 'model engineering'. Does this expression include the building and running of the clockwork or electrically-driven railway models of the smaller gauges — gauge '1', gauge '0', gauge '00' etc. — and what about the very small steam-driven models? And what about ship models? Can the building of miniature waterline ship models, perhaps only two or three inches long, or tiny model galleons in glass bottles, be accurately described as 'model engineering'?

It is difficult to answer these and similar questions without appearing to be dogmatic. But for the purposes of this book, I propose to regard the building (and running) of the following models as coming into the category of model engineering: model locomotives to scales of ½ inch to the foot and above, whether driven by steam, diesel, oil, petrol or electricity; traction- and similar engines (ploughing, agricultural or road haulage engines), road rollers, steam tractors and wagons; petrol engines of all kinds; marine and stationary engines, whether steam, petrol, diesel or gas, and also hot-air engines; power boats — especially their machinery; fairground and farmyard equipment; scientific and horological apparatus (e.g. microscopes, telescopes, clocks and so forth). Model engineers may also make their own equipment for the workshops, ranging from elaborate and complicated electrically-driven machines to simple hand tools, jigs or fixtures. These last cannot be called models, but are within the model engineer's scope.

A model of the U.S.S. Constitution by Bernard G. Phillips of Slough.

Some readers may also be surprised to see clocks included in my list. They may insist that making clocks can only be described as clock-making! They will say that this branch of craftsmanship cannot possibly be called model engineering. This is perfectly true; but the fact is that model engineers *do* build clocks, using much the same types of tools and machines as are used for locomotive or traction-engine models. Then again, many of the operations involved in clock-making are exactly the same as those used in other branches of model engineering, whether it is straightforward turning, milling, drilling, gear-cutting or soldering. And there is a great deal of highly skilled hand work involved.

Much of the craft of model engineering is very closely allied to full-size mechanical engineering; in fact there is no definite dividing line. The building of a 1½ in. scale model locomotive or a 3 in. scale model traction-engine may be regarded as model-making or it may be regarded as mechanical engineering on a small scale. The second description is perhaps nearer the mark and is one which will probably carry the greater appeal to the enthusiastic model engineer.

The fundamental laws of mechanics, the properties of steam, liquids and gases, the strengths and characteristics of engineering materials apply equally to the model engine as to the full-size engine developing thousands of horse-power. The design principles long established for the successful operation of the full-size machine apply with only minor modification to the miniature engine.

Some model engineers prefer to build their models to a well-known published design, with the confidence that, given reasonable workmanship, the resulting model will perform in a satisfactory manner. Many, however, are not content with this, but prefer to design their model themselves, much hard work on the drawing-board being required before metal is cut.

To design a successful model, even a simple single-cylinder steam-engine, is a much bigger undertaking than many people realize. It requires a good knowledge, not only of engine design, but of the correct use of the different materials, the strength of fastening devices, such as screws, bolts, rivets, and so on, and the best methods of carrying out the various machining and assembly operations when construction begins.

A "Bulle" electric clock made by W.C. Foster.

Locomotive Picture Gallery

LBSC acquired a full-size L.B.S.C.R. signal to put alongside his garden railway at Purley Oaks, Surrey. The locomotive standing in front is his 3½ in. gauge single "Grosvenor".

A beautiful glass-case model of the famous G.N.R. "Stirling Single", built by A.G. Peacock of Wroxham. It won both a Silver Medal and the Duke of Edinburgh Trophy at the Model Engineer Exhibition.

A model of the "Lion" built by L.A. Saxby of Hillingdon. This engine gained the Championship Cup at the 1970 Model Engineer Exhibition.

Another view of Mr. Peacock's fine ¾ in. scale Great Northern "Stirling Single".

The late LBSC was fascinated by the huge driving wheels of some of the early "single drivers". He built this unusual 2-2-2 for 3½ in. gauge, though it does not pretend to be a close replica of any of the Great Western engines.

A Bagnall contractor's locomotive with Hackworth valve gear built by H.C. Linck, seen at a Newport Exhibition. Note the wooden buffers designed to cope with wagons of different buffer heights.

A small industrial 0-4-0 saddle tank locomotive built by Peter Dupen, which gained a Championship Cup at the Model Engineer Exhibition.

Saddle tank locomotives were greatly used by collieries and private companies. This neat 5 in. gauge model was built by Mr. D. Oxland.

A very fine model of the famous Brighton "Terrier", to the author's "Boxhill" design. It is for 5 in. gauge. In spite of its diminutive size, this locomotive has hauled no less than seven passengers.

Believed to be the first successful 2½ in. gauge passenger-hauling coal-fired locomotive. LBSC's original "Ayesha".

At the "Battle of the Boilers", a locomotive trial held at the 1924 Model Engineer Exhibition, LBSC demonstrated the possibility of coal-firing in this small scale, showing a better performance than a big 3-cylinder spirit fired 2-8-2 locomotive designed by Henry Greenly and built by Bassett-Lowke Ltd. of Northampton.

ODE TO THE LONDON, BRIGHTON & SOUTH COAST RAILWAY

All things Brighton beautiful
Their locos great and small
Coaches weird and wonderful
The Brighton made them all.

The Gladstones and Atlantics
The gorgeous Southern Belle
The Terriers and the D.Is
The bogie tanks as well

And also Stroudley E tanks
The radials by R.J.
At any Brighton station
You'd see them any day.

And then there were the Scotchmen
Goods C.2s and C.3s
And later still the J tanks
(Such lovely engines these!)

Then lastly came the K class
The splendid 2-6-0s
The I.3s and the Baltics
None could compare with those.

Victoria to Brighton
In sixty minutes flat
They said that they could do it
They did! and that was that!

I'm glad I'd eyes to see them
And lips that I might tell
How clever was the Brighton
Who made them all so well.

(A.E. Rothon)

A fine model engineer's workshop in the West Country. Machines include lathe, vertical milling machine, drilling machine, arbor press, grinding machine, bending rolls etc. Note the models at the far end, and the "Atlantic" locomotive chassis on right.

On Workshops

IT ALWAYS amazes me how some of our model engineers manage to turn out superb work in quite primitive workshops. I have met those who quite cheerfully work away in tiny garden sheds, often at some distance from their house or flat, and often with rather primitive heating and lighting. More power to their elbows, but even when financial resources are limited, surely something can be done about the problem.

Where the domestic arrangements permit, the model engineer's workshop is best sited in the house itself, as apart from the much easier heating arrangements obtainable, one is not entirely cut off from the rest of the family. Those who are fortunate enough to be able to use a room in the house are, however, well advised to find some other place in which to do their soldering and brazing, as apart from the noise and heat involved, there is the danger of rust on valuable tools and models.

In some of the older houses, an attic or a basement may be pressed into service as the workshop. Neither is ideal. Attics tend to be too hot in summer and too cold in winter. It may be very difficult to carry the heavier equipment, lathes or other machines up to the attic, and if one has to do any heavy riveting work, there is bound to be friction with other members of the household or even with the neighbours, whatever precautions are taken to minimize the nuisance. Basements may be ideal for doing some of the heavier, noisier work, but they too have their drawbacks. They are often cold and damp and may have little or no natural lighting.

Some houses, especially those out in the country or in country villages, have brick-built outhouses, often built directly on to the house itself. These generally make excellent workshops. My own is of this type, and although very small, it has many advantages.

Then there are the various types of outdoor workshops, which may be of concrete, brick or timber construction, or a combination of any of these. It should be remembered that the professionally-built portable timber workshop is often of very light construction, to keep down the price, so is not ideal for model engineering purposes without considerable strengthening. Concrete or brick is perhaps to be preferred, but in either case, the walls should be lined with timber or one of the many kinds of hardboard now on the market. Roofs, too, should always be lined, as this assists greatly in conserving heat.

Those who build their own workshops will not need reminding that the necessary planning permission must first be sought from the appropriate local authority. Other important points are the provision of adequate natural lighting, ventilation under wooden floors, sound damp courses in brickwork, and the provision of suitable flooring material.

There is nothing worse than having to stand on bare concrete, which is terribly cold in winter, and the provision of a suitable floor covering is most important. Although linoleum is sometimes used, as it is nice to work upon, it is not ideal as metal swarf soon becomes embedded in it.

One solution to the problem is to use old carpets, which have become too worn to be much use in the house. Although such carpets quickly 'collect' swarf, they can be taken out and shaken at regular intervals. Apart from the fact that even old and worn carpets are pleasant to work upon, they have the advantage that they prevent, to a great extent, the annoying business of the model engineer taking metal swarf into the house on the underside of his shoes!

Benches

Most model engineers seem to make their own benches; perhaps this is because most of the ready-made benches seem to be on the flimsy side; apart from that, the home-made bench can be built to suit the available space, an important consideration. The home-made bench usually has legs of at least 3 in. square section, and this is none too heavy, while the top planks are generally 1½ in. thick. A good dodge to increase rigidity, especially near where the vice is arranged, is to bolt on heavy diagonal timbers, running from the top of the front leg to the bottom of the rear leg. Such diagonals will take care of the thrust of heavy filing and sawing.

The height of the bench is often regarded as a controversial matter, the usual test being that, when the vice is mounted, the upper surface of the vice jaws is level with the operator's elbow when in the position for filing etc. Thus the bench height will depend on the height of the model engineer himself, although it generally works out at between 2 ft. 9 in., and 3 ft.

A 'de-luxe' workshop bench might have a plywood covering on the top, plus a 'backboard'; if the heavy top planks are left uncovered, small parts will be continually slipping down the cracks between them. Some may prefer a linoleum covering rather than plywood, especially if

A very fine workshop in South Africa (Mr. A.S. Prescott of Johannesburg). Machines include drilling machines, lathes, vertical and horizontal milling machines, arbor press, bending rolls, etc. Below: Another view. Each machine has its own adjustable light. A Myford Super-7 can just be seen on the extreme left.

As this picture shows, there is no reason why the model engineer's workshop should not be a comfortable place to work in, or where one cannot entertain a friend with common interests.

much light precision work is carried out, but in this case, a plain colour should be chosen, otherwise it will be difficult to locate small parts which have dropped during assembly work!

A final point about workshop benches. Don't forget to allow a clear knee space underneath, so that the model engineer can sit down comfortably for marking out, light assembly work and so on.

Storage

The storage of small tools and accessories, materials, screws, rivets and so forth is worth careful consideration, otherwise much valuable time can be wasted in looking for things. It is often said that as model engineers are amateurs, the time they take on building their models, or whatever they actually do in their workshops, is of no importance. I don't agree with this. Most of us have severely limited time in our workshops, so that time wasted on non-productive jobs is really quite important, and should be kept to the minimum.

Plenty of wooden shelves around the workshop are always useful, and while precision measuring devices such as micrometers, dial test indicators, and of course try squares, dividers, calipers and similar items should as far as possible be kept under cover in chests of drawers or cupboards, those tools which are in constant use are best kept in open racks or in spring clips, where they can be found quickly. I keep all my files (except for needle files) and also hammers, screwdrivers, chisels, etc. in spring clips. I also keep a selection of taps which are in constant use, both second and plug taps, ready mounted in pin chucks and held vertically by pairs of nails driven into the outer edge of a shelf. The size of each tap is clearly painted on the edge of the shelf close to the tap, so that no time is lost in looking for the right one for any particular job.

Model engineers who are regular smokers often keep their screws, rivets, etc, in old tobacco tins, their contents being marked on the ends of the tins. My own preference is to use clear plastic containers with lids. These can be obtained in the convenient size of 6 in. x 4 in. x 1 in. deep, and divided into nine compartments. They are also very useful for temporary storage of the small parts when dismantling models for repairs or painting.

The biggest problem facing most of us in our workshops is the rusting of steel parts, machines, tools, etc. It is sometimes said that the way to avoid rust on tools is to use them! But of course this is not always possi-

A general view of the workshop of Mr. A.R. Casebrook of Wolverton. On the right is his Myford ML7 lathe, drilling machine is on the left, also the assembly bench and vice. Small tools are kept in drawers on the right.

ble and in any case there are some tools, jigs, and so on, which are only required very occasionally, for special jobs.

The question of lining of walls and roofs has already been mentioned; but perhaps the only certain way of preventing rust is the maintenance of a suitable temperature throughout the autumn and winter months. Gas- or oil-fired convector heaters should be avoided. A low-wattage electric heater is probably the best solution, preferably fitted with a thermostat to control the temperature between 60 and 65 degrees Fahrenheit.

Good work cannot be done without good lighting, and there is nothing to beat daylight for any kind of work whether on the bench or on a machine. But when artificial lighting is needed, perhaps the most satisfactory method is a blend of 'general' lighting and individual lamps on the bench and on each machine. For the former, fluorescent tube lighting is economical. On the bench, there is nothing to beat a fully flexible portable lamp of the well-known 'Terry' type. For machine lighting, many model engineers prefer a low-voltage system, using a transformer, though great care must be taken to avoid confusing the wiring of the low-voltage system with the high-voltage mains.

The late LBSC had a most comprehensive workshop, though it was somewhat cramped. In this picture can be seen the assembly bench, with parts of a locomotive tender. At the rear is a bending rolls.

On lathes and other machines

THE lathe has been described, and rightly I think, as the 'king' of machine tools. It is really astonishing how many different kinds of operation can be done on the lathe, given the necessary accessories of course.

It is not my intention, in this chapter, to tell readers how to use their lathe; there are several excellent books available which deal with this subject in great detail. I would however like to offer some suggestions to beginners as to the type of lathe most suited to their requirements; also to make some comments on small lathes in general.

Most model engineers' lathes seem to come between the 3 in. centre and the 5 in. centre size (6 in. — 10 in. swing). Let me say straight away, therefore, that the very small model makers' lathes on the market, generally of 1½ in. to 2½ in. centre height, although quite satisfactory for light work, as for instance in building gauge '0' and '1' model locomotives, are unsuitable for serious model engineering work.

It is generally agreed that a good deal of the work done by the model engineer on his lathe is really too big for the machine, so my advice has always been to try and obtain the largest and heaviest lathe (within reason!) that the pocket or accommodation will allow. By this I am not suggesting that the amateur requires a lathe of 6 in. centre height and 3 ft. between centres, unless of course he is thinking of building locomotives of 1½ in. scale or larger, or large traction-engine models.

Whatever size of lathe is chosen, sturdiness of construction in the bed, headstock and tailstock is the most important consideration. Before the 1939-45 War, there were many small centre lathes on the market which were anything but rigid. The position today is fortunately much better in this respect; although the choice is considerably more limited, those small centre lathes still on the market are generally of good design and construction.

If the enthusiast's pocket does not stretch as far as a suitable new machine, there is a good deal to be said for buying a second-hand lathe

— if possible from a fellow model engineer. It is a sad fact that where small centre lathes are used in professional workshops, they are not always treated with the respect that they deserve, so that the used amateur-owned machine is more likely to be the better buy.

One of the important points in amateur lathe purchase is whether to look for a machine with a gap bed. Most small American and some Continental centre lathes do not have a gap, so that a 5 in. machine will swing 10 in. diameter and no more. Many British machines of say 3½ in. centre height, provided with a gap in the bed immediately in front of the headstock, can swing a larger diameter than this.

The advocates of the 'straight' bed claim that the gap is a source of serious weakness. While this might be true if the gap-bed lathe is not properly designed, it may be said that most modern lathes are sufficiently strong at this point for all normal purposes.

Some earlier lathes were not provided with back gearing, some too had solid mandrels. Both types should be rejected out of hand. A good deal of the work done by the amateur requires a slow speed and ample torque, and both these are quite unobtainable in a small lathe without a back gear. There was however an exception to this some years ago, when the 'Exe' centre lathe was on the market. This machine had no back gear, but it was provided with a very large diameter pulley on the mandrel, which had much the same effect as the back gear, and was at the same time quiet and smooth in operation.

As to solid mandrels, these are an abomination and should be avoided at all costs as a very large proportion of amateur work calls either for the workpiece to be put through the mandrel, or for the mandrel to accommodate the usual taper shank of milling cutters, drills,

Among the larger, industrial, type of lathes sometimes used by model engineers, is this 6½ in. centre "Willson". It takes 36 in. between centres and has a full screw-cutting gearbox.

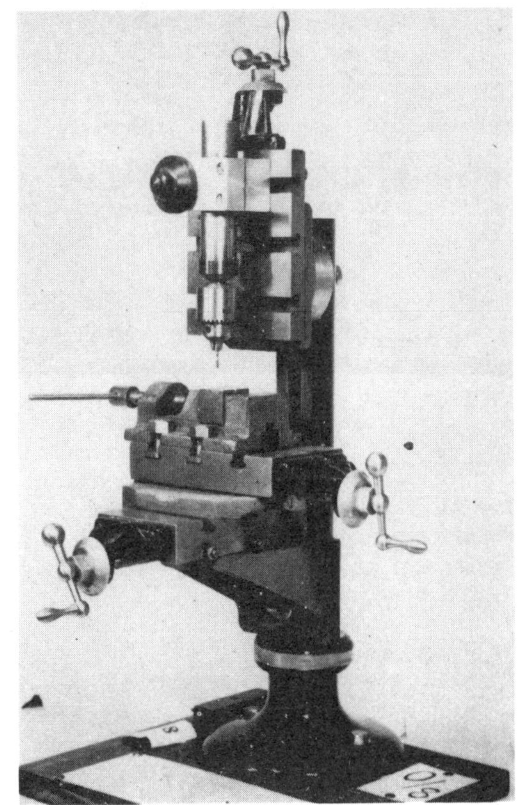

A fine 8 mm. watchmaker's type of lathe seen at a Model Engineer Exhibition. The four-jaw chuck seen fitted to the spindle would normally be considered too heavy for this type of lathe.

Mr. R. Cutler of South London made this ingenious vertical milling machine, utilising two Myford vertical slides and a Myford top-slide.

This 3 in. centre lathe was completely reconditioned by its owner Mr. Holder, who also made the fine accessories seen in the foreground.

reamers and so forth. Where possible, a lathe should be chosen with No. 2 Morse taper centres, this being a popular size among model engineers today. Lathes of 4 in. centre height or larger may have the mandrel bored for No. 3 Morse taper centres with advantage.

There has always been some controversy over the shape of the lathe bed 'ways'. While the raised V bed has certain advantages, mainly over the question of accuracy, it must be remembered that the old English flat bed is easier to manufacture accurately at a reasonable price, and it is also easier, from the amateur's viewpoint, to true up — by means of grinding or hand-scraping — should this become necessary after prolonged use.

Headstock bearings

The cheapest model engineer's lathe may be fitted with a half-split headstock bearing. This consists usually of a bronze bush, split longitudinally along one side and held in a circular housing in the lathe headstock casting, which is also split along the side. A bolt or bolts on the 'split' side are used to adjust for wear. This type of bearing cannot be called a good one; nevertheless many small lathes with these bearings, treated with care, have given quite good service.

Another simple headstock bearing is the 'split-parallel' type, where the adjustment is carried out by means of bolts on both sides of the spindle. The bearing proper is a split bush, generally of gunmetal or phosphor-bronze, though occasionally of cast iron, or steel with a white-metal lining, and packing shims are used for adjustment. This is quite a good bearing and not expensive to manufacture.

The more expensive lathes are generally fitted with headstock bearings of the taper type, or of the roller type. The roller bearing headstocks, incorporating 'pre-loaded' bearings, are generally of the tapered type, and in some lathes, use is made of a combination of ball and roller bearings, arranged so as to take up the end thrust of drilling and similar operations.

Toolholders

The subject of toolholders for the small lathe seems to be a highly controversial one. Many model engineers favour the four-way tool-

The Boxford ME. 10 centre lathe is of 5 in. centre height and was designed especially for the model engineer. It has raised vee bed and pre-loaded taper roller bearings in the headstock. A later model has a screw-cutting gearbox in place of the loose change wheels.

One of the smallest lathes manufactured is the Cowell 90 made by the Cowell Engineering Co. of Norwich. It is based on larger machines and has a compound slide and a full set of change wheels for screw cutting etc.

holder or turret. In this, the four tools are arranged at correct centre height, and by means of a locating device, the required tool can be brought into turning position when required, the other three tools being swung out of the way. According to the design and workmanship of the indexing device in such a turret, the tools can be brought back into position with a repeatable accuracy of anything from 0.0005 to 0.005 in.

Combined with a rear toolholder for parting off, the four-tool turret is useful for short run repetition work; but it has a disadvantage in that the three tools not in use are not necessarily out of the way, either of other parts of the lathe or of the work, or of the operator.

Recently, the so-called 'quick-change' toolholder has come into prominence, both on the larger industrial lathe and on the model engineer's lathe. In this, a solid 'body' is arranged to rotate on a vertical pin, as with the 4-tool turret, but the lathe tool is previously clamped in a separate holder which itself can be clamped accurately to the body by means of a cam and two Vs on the holder. A height-adjusting device is incorporated, so that any tool previously clamped in its holder, can be very quickly brought to correct centre height. Any number of holders may be used, holding tools of any shape and size within the capacity of the holder, such as left-hand, right-hand, parting, roughing, screw-cutting, boring, etc., and by moving a single clamping lever, the desired toolholder can be brought immediately into operation.

While there is no doubt that these elaborate toolholders have many advantages, the simple clamp type of toolholder generally supplied as

An ornamental turning lathe by Sibley. Ornamental turning is a highly specialised form of this work, and many materials other than metal are used by its exponents, such as hardwood, ivory etc. Photograph by Roy C. Hungerford A.R.P.S.

standard equipment with the cheaper lathes should not be despised. At least they are rigid, and the tool can be clamped directly down on the surface of the top-slide with the minimum of overhang. The difficulty of height adjustment can be overcome to some extent by a set of packing pieces of different thicknesses. These can be made from steel or brass flat strip and should be in steps of about 5 thou. If they are numbered clearly, the packing piece of the required thickness to bring any particular tool to centre height can be selected very quickly.

Earlier, I mentioned rear toolholders. These are most useful for parting off. Beginners nearly always have trouble in parting off, generally due to the tool digging in. The phenomenon of digging in is generally due to spring in the lathe spindle rather than to any lack of stiffness of the tool, which is more easily dealt with. It is also sometimes due to backlash in the cross-slide feed-screw. It is not easy to eliminate all backlash in the feed-screws, but it certainly helps if the cross-slide gib screws are adjusted so that this slide is rather stiffer to move than would be desirable for general turning.

Other points to watch are that the parting tool itself should be ground quite square at the tip, and of course not made too wide for the capacity and power of the lathe; that the tool has proper side clearance on *both* sides and that it is presented quite square to the work. It should be set to

The "Quorn" tool and cutter grinder, designed by Professor D.H. Chaddock of Quorn, Leics., well-known contributor to Model Engineer. A most useful machine capable of re-grinding milling cutters of all kinds, drills, reamers, taps, etc.

The small Unimat lathe has been described as the universal machine tool for the model maker who has not yet graduated to the larger and more powerful machines. Ideal for light turning, especially in brass and the light alloys, this picture shows the machine fitted with a handle to the mandrel by the author, to enable small iron castings to be dealt with.

cut very slightly below lathe centre height (or above, if mounted upside down at the rear). If these points are followed, the tool is mounted at the rear, and if parting steel, there is adequate lubrication, parting off should present no great difficulty.

Do not be misled by those who say that a parting tool should be ground off at an angle — so as to eliminate the 'pip' on the end of the work. This is all right in brass turning, but not when dealing with steels or phosphor-bronze. Any 'pip' can be easily removed, and it is much better to make sure of good parting off rather than worry about a 'pip'.

Lathe equipment

Most model engineer's centre lathes are supplied, as an essential part of the machine, with the faceplate, a pair of centres — generally one soft and one hardened — and a set of change wheels for screw-cutting, unless the machine is fitted with a gearbox. Thus before much general turning can be done, it is necessary to obtain various accessories. Perhaps the most important accessories (after the actual turning tools) are chucks; a 4-jaw independent and a 3-jaw self-centring chuck are essential. A good drill chuck with arbor to fit the tailstock is also invaluable.

I have often been amused to hear arguments over which type of chuck should be purchased first, if available funds will not stretch to all three.

This is like asking which should one have first, the horse or the cart! But if one is really up against it, I would plump for the 4-jaw independent chuck first, as we *can* hold round stock in this while we can't hold square or irregular stock, or castings, in the 3-jaw! Some lathe chucks are themselves threaded internally so that they can be screwed directly on to the lathe spindle, a separate backplate being eliminated. This is an advantage as 'overhang' is considerably reduced. If a separate chuck backplate cannot be avoided, it should not be made thicker than really necessary for strength.

Other lathe accessories that are extremely useful, in fact essential if full use is to be made of the machine, are angle plates, and 'dogs' for holding work to the faceplate, a vertical-slide with vice for light milling work, fixed and travelling steadies, lathe carriers, tailstock die-holders and knurling attachments. Collets may be regarded as a luxury, but nevertheless, most useful for accurate bar work.

To complete this list, there are several more specialized items of equipment that the advanced lathe user may eventually acquire. These include running centres, drilling and milling spindles, taper-turning and spherical-turning attachments, revolving tailstock turrets, etc.

Lathe drives

The modern tendency in lathe design is toward the self-contained unit, the motor and countershaft being attached to the lathe itself with short V belts between motor and countershaft and between countershaft and spindle. There is however a great deal to be said for mounting the motor separately, as less vibration is transmitted to the lathe. (Even those motors on flexible mountings vibrate a certain amount).

The advantages of the V belt are better grip and the fact that the belt may be much shorter than a flat belt. In addition, the V belt is endless and therefore there are no joining problems.

The flat belt is unsuitable for short drives, so that the pulleys must be a considerable distance apart if excessive slipping is to be avoided. Flat belt fasteners need not be a great problem; a quiet drive can be obtained by methods such as lacing, scarfing and cementing.

One advantage of flat belting is the ease with which it can be slipped from one pulley to another, especially useful where an overhead line shaft is used, where several machines are operated from the one countershaft.

Before leaving the subject of lathes, a word about the lathe stand. Most lathes today can be supplied with a suitable metal stand and drip tray. While these are very convenient, they should not be regarded as absolutely essential. A wooden stand can be quite satisfactory but it must be rigid, and the lathe must be bolted down with care, as uneven bolting can pull a light lathe out of true. However, a metal drip tray is certainly useful, especially if much steel turning is done, involving the use of 'suds' or other cutting oils.

This is the well-known Myford ML. 10 centre lathe, an ideal machine for the model engineer starting in this fascinating hobby. A full range of accessories is available for Myford lathes.

Myford Ltd. of Beeston, Nottingham have a high reputation for small centre lathes and other machine tools. This picture shows their dividing attachment mounted on their 3½ in. "Super-7" lathe.

Other machines

Next to the lathe, the drilling machine is the most important machine in the amateur's workshop. In fact, many model engineers start with a drilling machine before they purchase their first lathe. While the lathe can be used for many drilling operations, in fact it is sometimes much the better machine for the purpose, the drilling machine is certainly much more convenient for the majority of such operations. For the lighter types of model work, the 0-¼ in. capacity drilling machine suffices, but for the model locomotive and traction-engines and similar projects, a ½ in. capacity machine will be found almost essential.

The modern drilling machines of ½ in. capacity or larger are almost always fitted with self-contained motor drive, and these are excellent for the model engineer except for one feature — their lowest speed is nearly always much too fast, certainly too fast for reaming, and often too fast for the largest diameter drill that can be accommodated. Another annoying feature is that the built-in on-off switch is often

This picture shows the "Senior" vertical milling machine set up in the country workshop of well-known model engineer Tom Walshaw.

A motorised hacksaw machine to the Bowyer-Lowe design made by the late G.W. Wildy. It gained a Bronze Medal at one of the Model Engineer Exhibitions. A most useful machine to save much hard work in the amateur's shop.

placed on the left of the machine. This is bad design. The switch should always be placed on the right-hand side, so that the right hand can be used to operate it while the left hand retains a hold on the vice holding the work. But a foot-operated switch is even better!

I have a 0-½ in. 'Fobco' drilling machine in my own workshop. This is an excellent machine in many ways, though it has the switch in the wrong position, and the underside of the table is so shaped that it is very difficult indeed to use a clamp to hold the work directly down on the table. My 'Fobco' has a useful additional feature — fitted by its previous owner — the belt cover, which was originally held down by two hefty knurled nuts, is now hinged along one edge, so that it can be lifted instantaneously for changing the belt from one pair of pulleys to another. This I find is a great time-saver, as it took several precious moments to unscrew the original nuts and replace them every time a change of speed was required.

Electric hand drills have limited use in the home workshop. The single speed variety run too slow for the smaller drills and much too fast for the larger drills (¼ in. to ⅜ in.) Most of them seem to make a shocking noise! Which reminds me, no one should despise the simple hand drill. My Stanley comes in for quite a lot of use, not so much for actually drilling holes, but to 'spot through', so that the job can then be transferred to the drilling machine.

Milling machines

I think most model engineers would agree that if space and funds permit, the next machine to buy after the lathe and drilling machine is a milling machine. The question is — horizontal or vertical?

If the reader is a locomotive enthusiast, one who builds locomotives exclusively, the horizontal milling machine is probably the more useful of the two; but there are disadvantages. The cutters for horizontal machines are very expensive, especially those big side-and-face cutters, and they require sharpening quite frequently, and unless one possesses a tool and cutter grinder, this can be a nuisance as well as a considerable expense.

For general model engineering work, the vertical milling machine is probably the best bet. Its great advantage is that end mills are much cheaper than face or side-and-face cutters, and much easier to make, at least the simple 'slot-drill' (which is really an end mill with only two cutting lips) and counterbores should be within the capabilities of every model engineer.

Grinding machines

Very little lathe work can be done unless some kind of grinding machine is available, on which to shape or sharpen the lathe tools. The old hand grinder is not often seen in toolshops nowadays, but it was never very satisfactory anyway. Unfortunately the modern 6 in. double-ended grinding machine is a rather expensive item, and many of the commercial ones on the market today have very inadequate tool rests.

The model engineer should therefore give serious consideration to making his own grinding machine. This is not at all difficult as ball-bearing 'plummer blocks' can be obtained, and suitable V pulleys. For 6 in. diameter grinding wheels, a one-third h.p. 1450 r.p.m. electric motor is just about adequate, driving the wheels at double this speed. A one-half h.p. motor would be even better.

The amateur can then 'go to town' over the tool-rests, which can be made to tilt in both planes, so that the proper front and side clearances can be ground on the lathe tools. Incidentally, the idea seems to have got around that grinding on the side of grinding wheels is dangerous, but this is not so if reasonable care is taken.

It need hardly be added that metal washers should never be placed directly against grinding wheels, but thick cardboard washers should be put between the two. Furthermore, the wheels should be well guarded, as a 'burst' can be very nasty indeed. Some protection for the eyes in the form of a clear plastic shield is also highly desirable.

A model paddle steamer (P.S. *Lily*) in the Liverpool Libraries. Reproduced by permission of the Museums and Arts Committee, Liverpool.

Ship Models

A working model steam drifter built by Capt. W. Stuart, U.S.A.F.

A waterline model of R.M.V. *Hibernia*, built by Mr. A.C. Yeates.

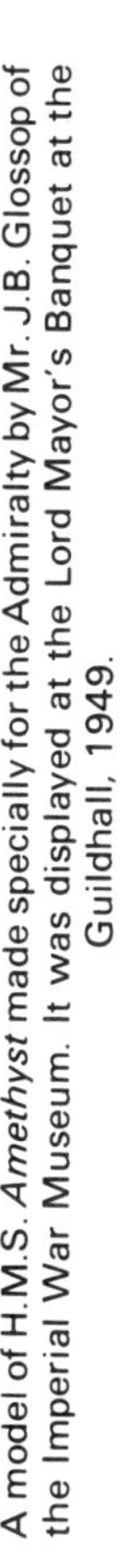

A model of H.M.S. *Amethyst* made specially for the Admiralty by Mr. J.B. Glossop of the Imperial War Museum. It was displayed at the Lord Mayor's Banquet at the Guildhall, 1949.

ON MOTION STUDY

The motion study expert
Looked in the Smithy door
And viewed with disapproval
The litter on the floor.
He frowned upon the Blacksmith
And timed his mighty swing
From the lifting of the hammer
To loud metallic ring.
He noted how his waistcoat
Was splitting at the seam
And many other details
Which showed a lack of scheme.

'My friend, you have no system'
He told the mighty man
'To maximize your throughput
You need an ordered plan.
We'll study all your movements
And draw up model rules
And redesign your bellows
And standardize your tools
We'll modernize your lighting
And regulate your hours
By statutory orders
And ministerial powers.'

The Smith leaned on his anvil
And spat upon his hands
Thus wasting the saliva
Secreted by his glands
He said he took it kindly
That anyone should be
So anxious to diminish
His inefficiency
And then, to show the expert
His quickness to respond
With minimum of effort
He threw him in the pond.

(Anon.)

On hand tools

MOST model engineers seem to build up their stock of hand tools in a somewhat haphazard fashion, buying additional tools as required for the job in hand. Undoubtedly, one of the most important items without which practically no model engineering work can be done is the bench vice, and this should be large and robust even if the work envisaged is not likely to be particularly heavy. One of 4 in. jaw width is none too large, and the coarsely serrated jaws fitted to most vices on purchase should be removed and ground smooth before use — perhaps a fellow model engineer may be able to carry out this grinding operation, otherwise it is well worth having it done professionally.

To hold delicate work, two strips of soft copper sheet may be utilized, these being bent to an angle section to prevent them slipping down, which can be rather irritating. A recent novelty on the market is the magnetic vice jaws, which have a facing of fibre. These of course stay in place all right, but they do tend to pick up steel swarf, which can be rather a nuisance; nevertheless I would not be without mine now.

Although most people taking up model engineering will almost certainly have a hammer or two, if these are purchased, always ask for the ball-pein type, as these will be found generally useful for riveting, planishing, etc. As for screwdrivers, although most of us have a good selection of these, how many of us keep them in really good condition? The secret of the screwdriver that doesn't keep slipping out of the screw head, is in the shape of the tip; it should be ground slightly hollow on both sides, easily done on a grinding wheel, and the end kept absolutely square.

Even if the novice model engineer has a drilling machine, the ordinary hand drill should not be despised — for many jobs it is indispensable — but it should be of good quality (look for lack of play in the bearings!). This leads me on to twist drills. Before the Metric system arrived, most of us invested in a set of number drills from No. 60 to No. 1, and if our pockets allowed, a set of letter drills, A to Z, plus a few fractional sizes — perhaps 15/32", ½ in. and ⅝. Then, if the type of work justified it (for making miniature injectors for instance!) a selection of very small drills between No. 76 and No. 62.

What to do about the Metric System? My advice to those who already have a selection of 'Imperial' size drills is to hang on to them and only buy the odd Metric drill when really necessary for a particular job. But for those who are just starting to equip a workshop, it may be safer to buy only Metric drills. The advocates of Metric dimensions will tell us that the Metric range of drills is more comprehensive and that the intervals between the sizes are more logical; but there isn't much in it.

I will have more to say about the Metric system when I come to screwing tackle.

I don't propose to say very much about files, except that the well-known dodge of painting their tang ends to denote which metals they should be used on should always be adopted. New files are right for use on brass and gunmetal, and these are painted yellow; partly worn files are painted black, for use on steel. We might also add another colour — green or blue — for files of fairly coarse cut that are too worn to be of much use on steel, but can still be used to clean up castings; or their ends can be ground to make scrapers!

The trade sell 'file cards' for the purpose of cleaning files which quickly become 'pinned' especially when working on light alloys. (For beginners' benefit, 'pinning' is the jamming of tiny particles of the metal being filed in the 'teeth' of the file.) But I have never found these of any use, and they tend to dull the cutting teeth of the file. A much better way of cleaning files is to use a strip of mild steel about ⅜ in. x ⅛ in. section, with its end filed to a chisel shape. Incidentally, although it is not often found in the model engineer's workshop, one of the most useful files is the 'pillar' file, a narrow flat file usually made in 6 in. lengths and with one or two 'safe edges'.

Another tool, and quite an inexpensive one, that is not often seen in the model engineer's workshop is the jeweller's or metal-piercing saw. These useful saws accept a great variety of blades, the thinnest being ideal for sawing very thin sheet metal, while the thicker blades are excellent for sawing thin-walled tubing, slotting very small screws and similar operations.

Screwing tackle.

One cannot do a great deal of model work without a reasonable selection of taps and dies. As a basic collection, I regard 4, 6, 8 and 10 B.A. (British Association) as almost indispensable, while some of the 'Model Engineer' threads — 5/32 in., 3/16 in., ¼ in. and 5/16 in. x 40T, 5/16 in. and ⅜ in. x 32T, and ⅜ in. x 26T are also most useful. For heavier work, tool-making etc. I like ¼ in., 5/16 in. and ⅜ in. B.S.F. It is very seldom that one needs to go above ⅜ in. diameter.

Once again, what about this new 'S.I. Metric' system that is now coming in? As I mentioned in connection with twist drills, those who are just

Another view of the late LBSC's workshop. He had plenty of hand tools, especially files and pliers!

starting a workshop would be well advised to go for the Metric screwing tackle, but for those who already have fairly comprehensive workshops, there is absolutely no need to scrap their 'Imperial' equipment.

It will be many years before the majority of model engineering drawings are only available with Metric dimensions, and even when this does come about, there should be no very great difficulty in building 'Metric' models with 'Imperial' equipment. To take just one example, there is no need at all to cut Metric threads, whatever type of model or workshop equipment is being made; the B.A. threads are very close to the S.I. Metric ones, and are excellent threads in their own right, at least up to 2 B.A.

For larger diameter threads, and for those that need to be screw cut in the lathe for greater pitch accuracy, the popular 40, 32, 26 and 20 pitches with Whitworth form will meet practically every eventuality.

As regards screws and nuts purchased 'outside', some of the model supply firms still stock these in B.A. sizes and in many cases with special small hexagon heads, which are ideal for model work.

My belief is that so long as there is a demand for such screwing tackle, at least one company will always be around to supply our needs. So 'Imperial' workshop owners — take heart, and hang on to your equipment!

The question which often arises is whether the high-speed ground-thread taps and dies which are now readily available are worth the extra cost. I believe they are, as they keep their cutting edges for so much longer, besides imparting a better finish to the work. But their high price may limit their numbers for many of us.

Storage of small tools, screws, rivets etc., often presents a problem for the model engineer. Here is how one enthusiast tackles it. Each drawer has a clear plastic front, so that the contents can be easily identified.

Those with screw-cutting lathes will be able to cut many of their threads, including those for which they have no taps and dies, in the lathe, and it should not be forgotten that threads produced in this way are more accurate.

Measuring instruments

Everyone must have a straightforward 12 in. steel rule, but how many of us have one of those most useful rules, a 12 in. flexible? These narrow flexible rules are far more useful on the lathe than the rigid wide type, though a 24 in. one of this type is essential for marking out the bigger jobs.

I hear a lot of arguments about whether one should buy a micrometer or a dial test indicator first, assuming of course that one cannot afford both together! But may I put forward an alternative? What about one of those dial-type vernier calipers, which can be obtained in the 6 in. size at quite a modest price. I have had one of these for several years now and find it most useful. It is almost as accurate as a micrometer and more versatile. However, the regulation 0-1 in. micrometer is certainly a most useful instrument and I should feel lost without it. I am very fortunate in also having 1-2 in. and 2-3 in. micrometers, but I would certainly not suggest that these larger micrometers are indispensable. In fact I seldom use mine.

The dial test indicator is a most useful instrument for lathe work, especially if it is mounted on a magnetic base; it has many uses, although its most frequent use will probably be for setting work to run true when held in a chuck or on the faceplate.

Reamers

Reamers are very useful for finishing holes to accurate dimensions, though in the larger diameters, apart from being very expensive, they are difficult to deal with, either by hand or in the lathe. I do not regard reamers as suitable for finishing cylinders of either steam-engines or petrol-engines, although for small steam-engines up to about ½ in. bore, they are generally successful. The important thing to remember about reamers is that they should only be expected to remove a few thou, anything more than .008 in. for reamers up to ½ in. diameter is likely to cause trouble — generally chatter marks or severe scoring. Spiral-fluted reamers are always to be preferred to straight-fluted, which tend to reproduce their flute pattern in the work. Taper pin reamers are most useful and are really essential if taper pins are used for securing collars, valve gear levers etc. to shafts. There are many kinds of adjustable reamers on the market, though the cheaper ones are none too reliable and I have never cared for them much.

Reamers should always be stored with great care. They should be kept in a drawer with partitions, so that they cannot touch one another;

to allow reamers to be jumbled up together is asking for trouble. It is also most important to keep rust away from reamers, and Shell 'Ensis' is most useful here. This can be sprayed on lightly, and is easily and quickly removed by petrol or paraffin when the reamer is required for use.

A final point on reamers; they should always be held in the lathe in such a way that they can find their own way into the work, in other words, they require to be allowed a little lateral 'float'. They should always be turned in the direction of cut and never reversed on withdrawal. Plenty of cutting oil should be used with reamers, except when reaming cast iron and plastics.

ON HEARING SMALL LOCOMOTIVES CALLED 'TIDDLERS'

I built a little loco
Its gauge is three point five;
It goes by coal and water
With steam both hot and live.

In days of dead electrics
And diesels of foul smell
I like my little loco
That steams and pulls so well.

There are many others like me
Who build these small iron steeds
In sizes nought to five inch
As satisfies their needs.

Our thoughts are somewhat bitter
When in a magazine
A sceptical contributor
Our offspring does demean.

Mr. C. Blair built this 3½ in. gauge L.M.S. 2-6-4 tank locomotive. The design named "Jubilee" was one of the first described by the author in the "Model Engineer". The author's original "Jubilee", after some running in this country, found a new home in Hamburg, W. Germany.

Though some of our small locos
Are less than one inch scale
He calls them Garden Tiddlers
And infers they are too frail.

We have no luxury workshops
With machinery replete
Our methods of construction
Our skills and pockets meet.

If he ever built a loco
In some back garden shed
In freezing cold in winter
On lathe with twisted bed.

And if, when finished, really worked
Though less than five inch gauge
And pulled perhaps a thousand pounds
He'd understand our rage!

(J. H. Owen)

Traction Engine Parade

This 4" scale Allchin leads a line up of miniature engines at a recent rally.

A model petrol-driven road-roller to the late Edgar Westbury's design.

A model traction engine built by Mr. D. Garrick seen on the footboard of a full-size Foster Compound traction engine.

A model Fowler class BB ploughing engine built by Colin R. Tyler of Tilehurst. It is to 2 inch scale.

One of the finest models of ploughing engines the author has seen. Built by R.C. Stone, and the winner of the Championship Cup at the Model Engineer Exhibition. Photograph by Anthony Beaumont.

The approximately 1/3 full-size Showman's engine built by Mr. Cyril Rose of Radcliffe-on-Trent. The working pressure is 130 p.s.i. and the engine generates at 24 volts. Construction started in 1948 and the engine was first steamed in 1962.

Al Park of Calgary, Canada built this 1½ in. scale Allchin traction engine to W.J. Hughes' design.

A fine Burrell traction engine built by Len Crane of Wolverhampton. Winner of the Duke of Edinburgh Trophy in 1972.

ROYAL CHESTER

How they do it in California, U.S.A. At the left is a Santa Fe 2-10-4 freight locomotive; in the centre is a Southern Pacific R.R. 4-8-4 Passenger engine, and on the right is a Baltimore & Ohio 4-6-4. The first two are to 1½ in. scale and work on 7½ in. gauge track. The third

On clubs and societies

THOSE who are unfamiliar with model engineering may be surprised to know that there are a very large number of clubs and societies devoted to the hobby, not only in this country, but in many countries abroad. The oldest and largest model engineering society is undoubtedly the Society of Model & Experimental Engineers, which was founded in 1898, shortly after the founding of the magazine *Model Engineer*, by the late Percival Marshall and a handful of enthusiasts. This society has permanent premises, with lecture room, library and workshop, in South London, and known as Marshall House. Meetings are held on Saturday afternoons throughout the year, except during the month of August, while the workshop, which is well equipped with several lathes, drilling machines, grinding machines, brazing tackle and so on, is available to members during the week.

The S.M.E.E. library contains a fine selection of textbooks and reference books on every possible engineering and model engineering subject, and bound volumes of the well-known model periodicals, and these are always available to members.

Today, most of the larger towns and cities in this country boast at least one thriving model engineering society, and many of these possess permanent railway tracks, generally for 3½ in. and 5 in. gauges, but sometimes for 2½ in. or 7¼ in. gauge in addition. The smaller gauge tracks are nearly always raised from the ground, on concrete or wooden posts, the rails themselves being either aluminum alloy section of flat-bottomed or 'Vignoles' type, or flat or angle steel, the latter sometimes being preferred on the grounds of better adhesion for the locomotives. The 7¼ in. gauge tracks are always laid direct on the ground, which allows the use of points and crossings, earthworks in the form of embankments and cuttings, according to the lie of the land, and other scenic effects.

Many of the model engineering societies own quite elaborate clubhouses. This one belongs to the Bedford Society, which owns a fine permanent track for locomotives.

At the International Model Locomotive Efficiency Competition. This is on the track of the Witney & West Oxfordshire Society's track in Blenheim Park. The engine is an "Eva May" designed by the late LBSC.

At the Andover Model Traction Engine Rally. On the right is a 2 in. scale traction engine built by Mr. Howell of Andover and on the left Mr. Baigent of Windlesham is with his 2 in. scale engine "Fair Lady".

Some of the model engineering societies tend to concentrate entirely on model steam locomotives, to the exclusion of all other types of models. Others specialize in model power boats, model sailing yachts or model aircraft. Other again cover a wider field, members building locomotives, traction-engines, stationary engines and marine engines, petrol-engines and equipment for their workshops.

A recent development has been the formation of associations of model engineering societies, the co-operation of such societies having many advantages, such as the possibility of staging exhibitions, the holding of model locomotive or traction-engine rallies, and the arrangement of economical insurance cover for their members. The principal such associations at the time of writing are the Southern Federation of Model Engineering Societies, the Northern Federation and the West Midlands Federation.

The individual model engineer will gain a great deal by joining his nearest society. He will invariably obtain sound advice on almost any subject connected with the hobby from the more knowledgeable members. He will be able to take part in the various club activities which include visits to places of engineering interest, and to exhibitions, regattas and rallies. He will also be able to enjoy talks by experts on the different branches of model work.

Southampton's railway at Riverside Park is well patronised. Easter Holiday crowds prove greater every year. Mr. William Perrett with his "Speedy" is hauling the train in the foreground.

At the International Model Locomotive Efficiency Competition. Well-known model engineer Norman Spink of Chesterfield examines the fire in his 5 in. gauge Great Western 4-4-0 locomotive "Gooch". Below: Mr. Weaver's 0-6-0 locomotive tackles the bank at Blenheim Park.

AT THE STEAM RALLY

My husband is a busy man,
Both in his work and leisure,
He's always planning things to do,
To give the people pleasure.

So in July, if all goes well,
A sight you'll see I'm glad to tell,
Here in this village, before your eyes,
You'll gaze in wonder and surprise.

A mammoth Rally is his aim,
A steam one (let me now explain)
Traction Engines, Showmans too,
Organs, Models, all on view.

Gallopers, for young and old,
Painted in blue and red and gold,
A dance at night, and barbecue too,
There'll be plenty there for you to do.

There'll be rides for the kiddies,
And hot-dogs to eat,
And seats for the elderly,
To take weight off their feet.

Two fine traction engine models, one to 3 in. scale and the engine in the foreground is to 2 in. scale.

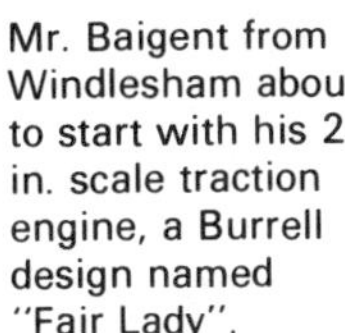

Mr. Baigent from Windlesham about to start with his 2 in. scale traction engine, a Burrell design named "Fair Lady".

There'll be steam-ploughing engines,
And chaff cutters too,
A horse-drawn steam fire engine,
A pleasure to view.

And don't forget on Sunday,
There's to be a Service as well,
We hope you'll help to sing the hymns,
And make the anthems swell.

The music from the Organ,
(Or maybe we'll have a band),
Will help you with the singing,
So come and give a hand.

There'll be hundreds of people,
From both near and far,
Coming on foot, by coaches and car,
Their fees for admission,
Will help us to pay,
For equipment now needed,
So that children can play —
In the field in our village,
Which at last now is ours,
And we hope will give pleasure,
For many long hours.

Mrs. M.R. Holloway

Vertical steam engine built by Anthony Beaumont of King's Lynn.

Stationary Engines

Another fine model beam engine built by Anthony Beaumont of King's Lynn.

A typical example of a model "grasshopper" beam engine. In this design, the beam is pivoted at one end, instead of at the centre.

A model of the Hick side rod steam engine in the collection of the late L.G. Bateman.

A twin cylinder entablature steam engine in the collection of the late L.G. Bateman.

A model of a Fairbairn winding engine generally built into the building housing them, hence the brickwork seen around the model. (Bateman)

A model of a Boulton & Watt beam engine built by H.J. Trevetts of Southbourne.

A model of a winding engine by E.D. Jenkins of Kettering.

A main and tail haulage engine to 1½ in. scale by J. Hollins of Stoke-on-Trent.

A model hot air engine built by W.D. Urwick of Malta. The machining was done on a universal lathe designed and built by Mr. Urwick.

A model of a table engine built by J.W. Ravenscroft of Leeds. It gained a Silver Medal at the 1970 Model Engineer Exhibition.

Marine Engines

A fine model triple-expansion marine engine. Built by Mr. G.E. Hartung of Gravesend.

A four-cylinder quadruple-expansion marine engine by Mr. C. Cole of Ampthill.

Possibly one of the finest model marine engines ever built, a side-lever paddle engine by the late Cdr. W.T. Barker. The original engines were built by Robert Napier in Glasgow in 1837. Scale ⅝" to 1 ft. Photograph by A.L. Sharp.

A fine triple-expansion marine engine built by Mr. E.V. Wilcox of Weaverham. It is complete with condenser, pumps, barring engine and thrust blocks. Awarded the Myford prize at the Model Engineer Exhibition.

On exhibitions

IN 1907 Percival Marshall staged an exhibition at the Royal Horticultural Hall in London. It was known as the Model Engineer Exhibition and against all expectations proved a great success. It has been held, with a few exceptions, once a year ever since.

At the first M.E. Exhibition, held over a period of five days in October of 1907, there was a long stream of visitors from all parts of the country, and even from abroad. There were those who had spent many years in engineering work of every kind. There were amateurs who had made models by the dozen, and amateurs who had never seen a model other than their own. Professors of engineering and other branches of applied science, service officers, doctors, dentists, schoolmasters, indeed people of every rank and calling, came, saw, admired, and maybe even went home determined to make a start in one of the finest hobbies ever, the hobby of model engineering.

In securing public appreciation and public recognition of the real quality and value of model engineering, this first exhibition worked wonders. It will probably be a surprise to the present day reader to know that at this 1907 Exhibition, there were no less than forty-one trade stands. Some of them bore famous names that are still familiar to the model engineer of 1977; for instance, Bassett-Lowke, Stuart Turner, Gamages, James Carson, Drummond Bros. Henry Milnes, Cassells.

There was a Model Engineer competition section, in which there were twenty-five entries. The first prize in class A, for engineering models, went to a Mr. John A. Barker for a Stuart vertical compound engine, and the second prize was awarded to a Mr. A. F. Hart, for a compound undertype engine. In class B, the first prize for electrical and scientific apparatus went to a Mr. Herbert Hildersley for a 160 watt dynamo. In class C, for ship models, the winner was a Mr. H.M. Savage for a model sea-going yacht.

In addition to the competition models and the trade stands, the

Official Catalogue cover for the 1911 *Model Engineer* Exhibition drawn by well-known artist W.E. Twining. Attractions included music by Herr Meny's Blue Viennese Band playing daily, and performing a *Grand March Model Engineer* specially composed by Herr Meny.

A picture of the Model Engineer Exhibition of 1936, at the Horticultural Hall. The stands of Messrs Buck & Ryan, the well-known tool merchants, and that of Messrs Bassett-Lowke of Northampton, can be seen, together with some of the competitors' exhibits.

Society of Model Engineers had an interesting display of their members' work, and two railway tracks, on which model steam and electric locomotives were demonstrated at frequent intervals. The Victoria Model Steamboat Club, the first club of its type in the country, showed a fine selection of model power boats, and a part sectioned model of the ill-fated *Lusitania*. The original full-size ship was, in a few years time, to fall a victim to a German torpedo.

The Model Engineer Exhibition of today is of course a very different affair, especially as regards the competition models, which have increased enormously both in numbers and variety. At the exhibition held at the New Conference Centre, Wembley, in 1977, there were no less than twenty-five model locomotives, in all scales from 4 mm. to the foot, to 11/16 in. to the foot. There were twelve items of model railway rolling stock, and as for the ship models, these numbered no less than eighty, ranging from miniatures (mainly below one foot in length), non-working steam and motor ships, power-driven boat models with power plants, non-working sailing ships, working yachts and sailing ships, and hydroplanes and speedboats.

The class for road vehicles was also well supported, with fourteen entries, mainly model traction-engines and similar engines. General engineering models included some fifteen stationary and marine steam-engines and internal combustion engines. Workshop entries numbered sixteen, non-working engineering scale models, scenic and representational models amounted to eight entries, and general craftsmanship, including clocks, numbered no less than forty-five entries. But this was not all, for there were some sixty exhibits made by juniors of both sexes, under the age of sixteen. Model aircraft too are now very popular, and four classes of these produced some seventy-three entries.

A recent development has been the introduction of special classes for military modelling, a hobby that has made great strides over the past five years or so.

One great difference between the Model Engineer Exhibition of 1977 and that of 1907 has been the very big increase in the awards made to the successful competitors. Possibly the most notable of these is the Duke of Edinburgh Trophy, a magnificent silver tankard presented as a result of the Duke's opening of the M.E. Exhibition of 1952. Any piece of model engineering work that has been awarded a championship cup or silver medal, or one of the cups donated by friends of the exhibition is eligible for the Duke of Edinburgh Trophy.

The next most important trophy is the Bradbury-Winter Memorial Challenge Cup, donated by Mrs. Bradbury-Winter and a number of

At the Model Engineer Exhibition, 1961, Mr. B. Grontos of Finland (left) the only foreign exhibitor at the Show, examines some of the ship models with the late S.L. Sheppard, a past Secretary of the Society of Model & Experimental Engineers.

Sir Miles Thomas (now Lord Thomas of Medmenham) opened the 1961 Model Engineer Exhibition at the Central Hall, London. He is seen here, left, examining a model of the "Vulcan" with the late Edgar T. Westbury, the Chief Judge at the Exhibition.

Laurie Lawrence drives the "M.A.P. Special" at the Model Engineer Exhibition at the Seymour Hall, and gives a ride to six members of the staff. The track is provided by the Society of Model & Experimental Engineers.

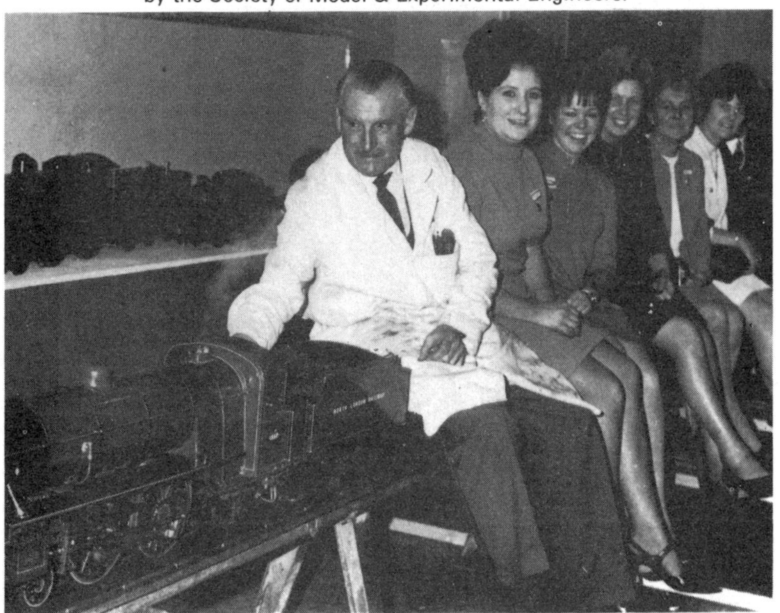

This is the Duke of Edinburgh Trophy, awarded in honour of the visit to the Model Engineer Exhibition of H.R.H. The Duke of Edinburgh. It is awarded for the best model of any kind that received a Championship Cup or Silver Medal (the two highest "class" awards) at the preceding exhibition.

At the Model Engineer Exhibition, several fine trophies are awarded in the different classes. The principal classes are for steam locomotives, small gauge railway models, traction engines, stationary and marine and petrol engines, and ship models of all kinds.

The Myford stand at one of the Model Engineer Exhibitions held at the Seymour Hall, London. Myford Ltd. of Nottingham are one of the leading manufacturers of small lathes and other machines.

friends in memory of the late Dr. Bradbury-Winter, one of the finest of the early model engineers, and the builder of the famous *Silver Rocket* which is now in the care of the Institute of Mechanical Engineers. This trophy is awarded for 'the most outstanding example of amateur craftsmanship'.

The Crebbin Memorial Cup was donated by friends of the late J. C. Crebbin, another outstanding model engineer, and is awarded for the best entry in the Locomotive, General Engineering and Mechanically-propelled Road Vehicle classes.

Other important trophies include the Aveling-Barford Trophy, presented by the directors of Aveling-Barford Ltd., as the annual award for the best working model traction-engine, road roller or steam wagon.

The Bowyer-Lowe Challenge Cup was donated by the late A. E. Bowyer-Lowe and is awarded for the best example of craftsmanship and design in the Tools and Workshop Appliances class.

The Bristol Challenge Cup was donated by the Bristol Aircraft Company and is awarded annually for the best model of a Bristol aircraft.

The J. N. Maskelyne Memorial Trophy, awarded in memory of a fine model engineer, railway enthusiast and draughtsman, is for the model locomotive, in any scale, which does most to promote fidelity of design.

In 1961, the Model Engineer Exhibition was held at the Central Hall, Westminster. This picture shows the Model Engineer Workshop Stand, with members of the Society of Model and Experimental Engineers answering technical queries.

The L.B.S.C. Memorial Bowl is a handsome silver trophy presented by public subscription by admirers of the late 'L.B.S.C.' (or 'Curly' Lawrence) and it is awarded to the best locomotive to one of L.B.S.C.'s designs, judged partly on workmanship and partly on performance, which is judged by putting the entries into steam on the track of the Society of Model & Experimental Engineers.

Among trophies awarded in the ship modelling classes, we should mention the H. V. Evans Trophy which was donated by the Thames Shiplovers' Society in honour of its founder, H. V. Evans. It is awarded annually at the discretion of the judges for research and/or presentation of a ship model.

The Maltby Trophy was donated by Mr. M. Maltby to be open for competition by Ship Model Societies exhibiting at the Model Engineer Exhibition.

The Maze Challenge Cup was donated by Sir Frederick Maze K.C.M.G., K.B.E., and is awarded annually for the best model of a sailing ship (pre-1820) or of an Oriental sailing craft of any period.

The Willis Challenge Cup was donated by Mrs. Willis, in memory of her husband, and is awarded annually for the best exhibit in the Hydroplane and Speedboat section.

The Exide and Drydex Challenge Cup was donated by the Directors of Chloride Batteries Ltd., and is awarded at the discretion of the judges for the best model incorporating a battery entered in any section.

The Edgar Westbury Memorial Challenge Trophy was presented by the Society of Model & Experimental Engineers and is awarded annually for originality and soundness of design in any type of prime mover, working plant in a marine craft, or fidelity to prototype of a working model i.c. engine.

For the small-gauge railway enthusiast, there is the Model Railways Bowl, presented by the proprietors of *Model Railways* for the best small-gauge railway exhibit.

The H. C. Wheat Challenge Cup was presented by his son in memory of the late H. C. Wheat, and is awarded annually for the best '0' gauge locomotive exhibited.

The New Zealand Cup was presented by the Model Societies of New Zealand and is awarded for the best model of a locomotive based on a design by L.B.S.C.

Finally there is the U.S.S.R. Central Marine Club Challenge Cup, donated by the Council of the Central Marine Club, U.S.S.R. and is awarded annually to the club whose exhibit represents the best achievement in marine model engineering.

In addition to the trophies already mentioned, a number of championship cups are awarded, where class numbers justify, to become the permanent property of the winners. Class trophies are awarded in the following classes: Locomotives, Steam or Motor Vessels, Sailing Ships or Yachts, Ship Miniatures, Mechanically-propelled Road Vehicles, General Engineering Models and Aircraft Models.

A new development at the 1977 Model Engineer Exhibition was a competition for model hot-air engines, which attracted a very large number of entries.

But it must not be thought that the London Model Engineer Exhibition is the only exhibition of its kind, even though in many ways it is unique. Several of the provincial model engineering societies hold exhibitions of their own, and in recent years there have been successful exhibitions at Bristol, Birmingham, Stockport, Sheffield, Cheltenham and many other centres. In addition to static exhibits of models, there are generally working models of stationary engines, small gauge working layouts and 3½ in.-7¼ in. gauge passenger-carrying tracks.

"Yes . . . it would make it easier to clean . . . however, how would it hold up under pressure?"

Reprinted from "Industrial Maintenance and Plant Operation" Philadelphia, U.S.A.

Clocks

A fine striking bracket clock with verge and crown wheel escapement made by the late J.C. Stevens of Ealing.

This picture shows part of the movement of a musical clock made by Claude B. Reeve.

The electro-magnetic engine, invented by Professor A. Pacinotti at Pisa in 1860. As a motor it runs on 12 volts, 2 amps, D.C. As a dynamo it gives 3 volts. The model is a full-size copy of the original by Dr. Paolo Rizzardi of Bologna.

A one-year spring and fusee bracket clock with perpetual calendar made by Claude B. Reeve of Hastings.

A quartz crystal timepiece made by Dr. A. Smith, built with simple hand tools, except for the motor and electronic components.

A skeleton regulator clock made by Claude B. Reeve of Hastings.

FAREWELL TO STEAM

Farewell steam locomotive friend
Your age of glory at its end.
A hundred years or more has been
Your presence on our daily scene.

From locomotive number one
To Evening Star your course has run;
Through more years than mere man's allowed
You've held your place supreme and proud.

With hiss of steam and plume of smoke
To us you lived, you breathed, you spoke,
A living thing of steel and brass
We'd travel miles to see you pass.

Conceived by men like Webb and Dean
Whose like again will ne'er be seen,
Gresley, Collett, Drummond brothers,
Churchward, Stanier, Hughes and others.

Ages of men have fashioned you
In towns named Ashford, Swindon, Crewe.
Draughtsmen's skill and craftsmen's art
Have carefully shaped your every part.

Clad in garb of every hue —
Bright Midland red and Garter blue,
From shades of green that looked so gay
To unlined black and wartime grey.

In many guises you were seen,
From sleek express with paint agleam
To dirty little saddle tank,
Shunting trucks with bump and clank.

Remembered are the names you bore
Of peaceful Saints and Men of War,
Earls and Dukes and Ladies Royal,
Racehorses and Regiments loyal.

We loved you first when as mere boys
Your likeness was amongst our toys,
And then in later life to find
We could not cast you from our mind.

Some of the S.M.E.E. visitors to the track in 1960. Left to right can be seen Messrs Mace, White, Townsend, Vickerage, Storey, Ewins, Carter, Hatherill, Brock, Bannister, Wildy, Rogers and Peck. Photo by Ann Hatherill (nee Carter)

You took us off to work each day
To funerals sad and outings gay;
Carried our fathers off to war,
Not always brought them back once more.

Took business men on hurried trips
Or sailors back to rejoin ships,
And 'neath romantic summer moons
Took newlyweds on honeymoons.

You hauled the wealth of this fair land
In loads of coal and iron and sand,
Machinery and milk and cream,
Transported all by power of steam.

Pounding northward over Shap,
Or speeding west through Goring gap,
You travelled on in every weather
From city's grime to Highland heather.

In lazy sunlit summer days
With exhaust steam the merest haze,
Or in the winter cold and bare
Your breath hung solid in chill air.

You never slaughtered, hurt or maimed
The thousands that road transport claimed,
But days you erred remembered still
At Sevenoaks and Quintinshill.

Leave us some mem'ries ere you go,
Night sky illumed by firebox glow,
The pungent smell of smoke and oil
And shovel's clank as firemen toil.

The age of steam draws to its end;
No more will locomotives send
A plume of smoke high in the air —
A sight that made men stand and stare.

As heavenward your exhaust curled
You changed the history of the World;
And so we say 'Well done — farewell!'
Your reign is o'er, you served us well!

(John Whittington)

On personalities

FOR my last chapter, I propose to tell the reader something about the many interesting personalities I have known during my model engineering career. Fear not old colleague, I shall say nothing that might make your ears burn! Which reminds me, isn't it strange how difficult it is to tell someone whom we like how we enjoy their company or how highly we think of them in one way or another. Yet immediately someone we know passes on to a Better Place, we all fall over ourselves to say the nicest possible things about him! But here at least I can let myself go, and recall only the nicest things about those I shall discuss — and you know, even those whom we could never bring ourselves to like in any way had their good points!

I have always regarded it as one of my greatest misfortunes never to have known the late Percival Marshall. I saw him at some of the earlier Model Engineer Exhibitions of course, but being only in my teens, never dared to address the great man. Yes, there is no doubt in my mind about this: he was a Great Man. To found a magazine so early in life, when the hobby was very much in its infancy, and against the advice of most of his friends; a magazine that is still going strong after seventy-eight years: in addition, to form the first model engineering society in the world, a society that is also still thriving all these years later! — undoubtedly these were very great achievements.

Another great model engineer whom I never knew personally, though I spoke to him occasionally at exhibitions, was the late Henry Greenly. Greenly was rather a controversial figure, but there is no doubt that he was a great pioneer and a fine model and 'full-size' engineer. It is very easy to pick holes in his designs, some fifty years later, but at the time Greenly was writing for the *Model Engineer* and other magazines, the hobby of model engineering was only just finding its feet.

The late Percival Marshall. Founder of Percival Marshall & Co. Ltd. and the "Model Engineer" in 1898. Also co-founder of the Society of Model & Experimental Engineers.

The present-day model engineer is inclined to forget that in Greenly's day, silver-soldering and brazing were regarded as too difficult for the amateur; there was very little brazing equipment to be had, no bottled gas blowpipes were on the market, and knowledge about hard soldering generally was limited. Is it surprising then that Greenly's boilers were generally made with gunmetal castings, screwed tubes and soft solder?

It is the fashion today to decry Greenly's classic work *Model Steam Locomotives*, but it is seldom remembered that this book was first published in 1922, long before even 'L.B.S.C.' had been heard of by the

A rare and important photograph. It shows the famous "Challenger", a 2½ in. gauge spirit-fired 2-8-2 locomotive built by Bassett-Lowke Ltd. to Henry Greenly's design as a challenge to LBSC, who ran his 2½ in. gauge coal-fired "Atlantic". Mr. W.J. Bassett-Lowke is seen behind the track, while Henry Greenly is next to the driver. The picture was taken at the 1924 Model Engineer Exhibition.

model locomotive enthusiast. Although it has been revised more than once, it is hardly surprising that this famous book has become a little out of date. Nevertheless, it still contains much useful information, and no model engineer of the mid-1970s need feel ashamed to delve into its pages, as indeed this author does on many occasions!

Let us not forget too, Greenly's many other books, almost as well known as *Model Steam Locomotives*, works on model railways of the smaller gauges, on electric railways, on signalling, on trackwork, and on model engineering in general.

In addition to his work on model engineering and model railways, Greenly was a fully-qualified professional engineer and did much useful work for such organizations as the Southern Railway. His career in connection with locomotives commenced about 1899. His first 15 in. gauge engine, the famous *Little Giant*, was of the 'Atlantic' wheel arrangement, and was built by Bassett-Lowke; this was the pioneer of British miniature railway locomotives and it was followed a few years later by the equally famous *Sans Pariel*, which saw considerable service on the Ravenglass and Eskdale Railway in Cumberland.

Henry Greenly's 'Pacific' *Colossus* was contemporary with the famous *Great Bear* of the Great Western Railway, these two being the only 'Pacifics' built in Britain up to 1922. Then followed Greenly's well-known one-third full-size 'Pacifics' and 4-8-2s for the Romney, Hythe and Dymchurch Railway in Kent. I wonder how many readers know that Greenly was the first to introduce both the 4-8-2 and 2-8-2 type of locomotive to this country? The 2-8-2 was the famous *River Esk*, designed by Greenly for the Ravenglass and Eskdale, while the 4-8-2 type was, as just mentioned, built for the R.H. & D.R.

River Esk, incidentally, also became famous as the first engine in Britain, of any gauge, to use the Lentz poppet valve gear. This was made by Davey Paxman of Colchester to M. Lentz's designs, but was not altogether successful, and was later replaced by Walschaerts valve gear.

I think that Greenly's locomotives for the Romney, Hythe and Dymchurch Railway represent his greatest achievement, certainly in the locomotive design field. By this time, 1924, he had profited by his experience at Eskdale and other lines, and as a result, specified substantial bearings and working parts for the Romney engines, and have they lasted! Of course they have had new boilers now, complete with superheaters, and no doubt many of their working parts have been overhauled or renewed; nevertheless, they are still basically Greenly engines. In the summer of 1968, I paid a long-promised and long-deferred second visit to Romney, and was pleasantly surprised at the superb condition of all the locomotives there, the Greenly engines and the others. In spotless condition, their regular exhaust beats betrayed their excellent maintenance.

One criticism of Greenly that is occasionally voiced is that he 'never actually built a locomotive with his own hands'. I cannot say whether this is strictly true or not, but even if it was so, did Churchward, Stanier,

Lord Northesk was a great enthusiast for the model steam locomotive and is seen here driving a 3½ in. gauge Great Western 4-4-2 locomotive built by Bob Gale. Lord Northesk was formerly president of the Society of Model & Experimental Engineers.

Gresley or any of the other great 'full-size' designers actually build their engines themselves? Of course not!

If I did not know Henry Greenly personally, I knew one of his chief mechanics, A. P. Campbell, very well. In fact Campbell joined my old model engineering firm some twenty-one years ago, and worked for me intermittently (he was not then in good health) for about eighteen months. Campbell was a brilliant craftsman of the old school. Give him a few hand tools and an old lathe of doubtful accuracy and he could turn out almost anything. Readers may have seen samples of his locomotive work in two beautiful Great Western Railway models, a 'Castle' four-cylinder 4-6-0, which is in the Science Museum, South Kensington, London, and a Dean 'Single' 4-2-2 which is (at the time of writing this) in store, but formerly used to grace the station entrance to the Great Western Royal Hotel, Paddington. If I remember aright, both these locomotives are built to 1 inch to the foot scale, for 4¾ in. gauge, which was a popular gauge at the time.

A. P. Campbell was one of those craftsmen who just could not bring himself to turn out what we used to call, in the trade, a 'cheap' job. It was no good my saying that the particular customer concerned only wanted a simple job at a low price, and if he were to put his usual standard of work into the model, we should both be badly out of pocket! No! Arguing was useless, the model had to be made to the usual Campbell standard of workmanship or not at all. Which reminds me! Campbell used to amuse me with reminiscences of his days working for Henry Greenly. Apparently Greenly had the same problem as I had with Campbell. And according to Campbell, Greenly used (on innumerable occasions) to wag his finger at him and tell him that it was pointless 'casting pearls before swine'! I only hope that Greenly's customers never overheard such a remark, or they might have very quickly transferred their custom elsewhere!

And now for that remarkable character L. Lawrence, alias 'L.B.S.C.' alias Curly, with his 'Inspector Meticulous', 'Pat', 'Uncle Jim', Driver Irwin, 'Bro. Wholesale', Arabs, fellow-conspirators and all!

It is only natural that I should go straight from Greenly to L.B.S.C., for these two pioneers were as different as chalk from cheese.

I first heard of L.B.S.C. just before the 1939-45 War. Late in 1939 I left employment with a model-making company in Regent Street, London, and was waiting to join His Majesty's Royal Navy. Having even then a fairly well equipped workshop, I was toying with the idea of producing a 3½ in. gauge G.W.R. tank locomotive, with the idea of putting it on the market after the war. This was to be one of the 61xx class, or

Sir Ronald East, a well-known Australian Civil Engineer, driving his "Boxhill".

The late J.N. Maskelyne, grandson of the co-founder of the once-famous magical entertainments at St. George's Hall. He is seen when giving a broadcast on the hobby of building model locomotives, and is holding a Gauge One steam locomotive built by the author to Mr. Maskelyne's design.

'large Prairie' as they were called, a type of locomotive which up to that time had not found very much favour with model engineers.

Although I did not know it at the time, L.B.S.C. had been contributing regular articles on model steam locomotive construction in the magazine *Model Engineer* for many years before this and was very well established as the authority on the subject. I therefore decided to write to him, care of the *Model Engineer*, asking for his advice on my locomotive project. Almost immediately, I received a long letter back, giving much useful information as to suitable cylinder sizes, valve gear details and notes on constructing a suitable 'Swindon' boiler. L.B.S.C. was most insistent that I should specify large ports and valves, warning me not to have anything to do with what he described as the tiny pin-holes preferred (so he claimed) by Greenly! Needless to say, I took his advice on this point.

It was shortly after this correspondence that I was able to read some of the back numbers of *Model Engineer* and here I quickly fell under the spell of L.B.S.C. It seemed that he was first noticed by readers of *Model Engineer* when he contributed some letters on model steam locomotive design and construction from November 1921 onwards. A great argument developed about this time as to the possibility of coal-firing for steam locomotives of 2½ in. to 3½ in. gauge, and also whether 2½ in. gauge engines could (or should!) be used for hauling 'live' passengers.

Up to about 1923, nearly all the 2½ in. gauge locomotives built were fired by methylated spirit or paraffin, and they generally had simple water-tube boilers, though a few builders had managed to construct locomotive type boilers using rivets and soft solder for caulking. L.B.S.C. would have none of this; he argued that proper locomotive type boilers fired by steam coal were much more efficient than those fired by spirit or paraffin, also that they were closer to full-size practice and more interesting to operate.

As to passenger-hauling, he claimed that even a medium-sized ½ in. scale express passenger type of locomotive would easily haul a live passenger, and to prove his point, set about building an L.B.S.C.R. 4-4-2 'Atlantic' to this scale. This model he named *Ayesha* after a character in

Author Nevill Shute with Gilbert Harding. Nevill Shute was a keen model engineer, and the central character of his book *Trustee From the Toolroom* was based on *Model Engineer*'s Edgar Westbury.

The late Claude B. Reeve, an expert amateur clock-maker, winner of innumerable prizes and awards, seen here with one of his clocks, an 8-day gravity escapement regulator.

one of Rider Haggard's novels, and it was to gain world-wide fame in the so-called 'Battle of the Boilers'.

Messrs Bassett-Lowke Ltd., the famous Northampton firm of model builders, was at this time producing a range of spirit-fired model steam locomotives for 2½ in. gauge, and to some extent for 3¼ and 3½ in. gauges, and no doubt they feared that L.B.S.C.'s claims for the coal-fired model might have an adverse effect on the sales of their own specialities. The result was that Bassett-Lowkes commissioned Henry Greenly (who at that time was acting as consulting engineer to the company) to design a large and powerful 2½ in. gauge spirit-fired locomotive, with which they could challenge L.B.S.C.'s claims. The model was quickly built by Bassett-Lowkes and was aptly named *Challenger*.

Challenger proved to be a three-cylinder 2-8-2 goods engine of free-lance appearance, and Bassett-Lowkes then challenged L.B.S.C. to a contest to be held at the 1924 Model Engineer Exhibition. L.B.S.C. responded to the challenge with alacrity and brought up his little Atlantic *Ayesha*.

It is interesting to compare the dimensions of these two very different locomotives. *Challenger* had three cylinders, ⅝ in. bore by 1¼ in. stroke, with the valves operated by slip-eccentric valve gear. The boiler was of the wide firebox type, the barrel being 3¼ in. diameter at the smokebox end and 3½ in. diameter at the throatplate. It was fired by a methylated spirit lamp having no less than ten ⅝ in. diameter wick tubes. The weight of the locomotive was 28 lbs., and ballast of 12 lbs. of lead were added to give greater adhesion.

Ayesha was fitted with two cylinders, 11/16 in. bore by 1 11/16 in. stroke, working pressure 80 p.s.i. Weight in working order 18 lbs. 10 ozs. But the most important difference between the two locomotives was that *Ayesha* had a coal-fired locomotive type boiler, whereas *Challenger* was fitted with a water-tube boiler. Mr. W. J. Bassett-Lowke claimed that the chief reason behind the building of *Challenger* was to show that a spirit-fired 2½ in. gauge locomotive could haul a live passenger and could 'perform continuously'.

The trials were held on the track of the Society of Model and Experimental Engineers, and *Challenger* was steamed first. Starting from all cold, a pressure of 100 p.s.i. was raised in nine minutes, and after a brief warming-up run, a fifteen minute run was made up and down the track, during which time twentythree trips were made. The fuel consumption during the run was stated to be one pint of spirit. The average drawbar pull registered was 4.5 lbs.

L.B.S.C. then steamed his 'Atlantic'. Working pressure was raised from all cold in 7½ minutes. At the end of fourteen minutes fifty seconds running, twenty-two and a half had been made, with L.B.S.C. himself at the regulator. The drawbar pull (average) was approximately the same as for *Challenger*.

In a letter to the *Model Engineer* after the trials, Mr. W. J. Bassett-Lowke seemed to be under the impression that the *Challenger* had been the 'winner', and he stated that in runs later on during the course of the Exhibition, the *Challenger* hauled a load of 476 lbs. In reply to this, L.B.S.C. wrote to say that in trials on his own line, his *Ayesha* had hauled no less than 563 lbs., a feat that he said was witnessed by two distinguished model engineers, Mr. W. B. Hart and Mr. J. Crebbin. When it is borne in mind that the *Challenger* was a very heavy goods engine, with three cylinders, small driving wheels, and a high working pressure, it is clear that L.B.S.C.'s locomotive was in fact very much superior. Added to this, the *Challenger* was very heavy on fuel, burning one pint in fifteen minutes' running.

The late Edward Bowness. Formerly Editor of "Ships & Ship Models".

The late Edgar T. Westbury. A distinguished model engineer, a regular contributor to "Model Engineer" for many years, author of many books and an expert on the small petrol engine.

Unfortunately, the correspondence which followed this interesting trial eventually became somewhat acrimonious, but L.B.S.C. had the last laugh, for *Challenger* did not stay in its original guise for long, as a later owner had her rebuilt and fitted with a coal-fired boiler. Whether it performed very much better after this, the author is unable to say, as nothing more seemed to come to light about this unfortunate locomotive.

The Duke of Edinburgh at the Model Engineer Exhibition. He is examining some equipment on the stand of the Society of Model & Experimental Engineers. In the centre of the picture is Professor D.H Chaddock and behind the Duke is the late Jim Crebbin, a well-known model engineer and great enthusiast for model compound locomotives.

The upshot of all the activity at that 1924 Exhibition was that L.B.S.C. was asked by Percival Marshall, the editor of *Model Engineer* to contribute regular articles on small steam locomotive design and construction for that magazine, and thus began the remarkable series for which L.B.S.C. became justly renowned.

Little seems to be known about L.B.S.C.'s early life, except that he was born into humble conditions and that at an early age, he joined the London, Brighton and South Coast Railway, working his way up through cleaner, fireman and driver. After he left the railway, he spent a short time in the U.S.A., and it was soon after his return to this country that the events I have just described took place. L.B.S.C. was to spend the rest of his life building, overhauling and repairing model locomotives, and writing about them. His regular articles published in the *Model Engineer* appeared from 1924 until his death in November 1967, apart from the period between 1959 to 1966, during which he was in disagreement with the editorial staff of that magazine.

Much has been written about L.B.S.C.'s remarkable workshop, which was so full of machine tools that one could hardly turn around without bumping one's shoulders against a lathe or a milling machine! His principal machines were a Milnes 3¾ in. centre lathe (a very good lathe), a Myford Super-7, and a Boley precision lathe, which he used for making injectors, small steam fittings and so on. His collection of hand tools and measuring instruments would have been the envy of many model engineers. When I visited him in 1964, I remember being struck by the absence of any kind of assembly area, a short bench being occupied by a 3½ in. gauge locomotive which he had completed. In an upstairs room, sparsely furnished, but decorated with many pictures of locomotives, L.B.S.C. had a small drawing-board, on which his designs were prepared for publication. Perhaps his most remarkable gift was that he was able to build quite an elaborate locomotive without any proper drawings at all, apart from a few rough sketches made in pencil on an odd scrap of paper. As he was able to visualize the finished locomotive, he had no need for drawings.

At his home in Purley Oaks (South Croydon), he had built a continuous raised track, for 2½ in. and 3½ in. gauge, and had installed beside the line a full-size ex-L.B.S.C.R. signal, which could be seen quite easily from the 4 ft. 8½ in. line which passed close by. On this track, L.B.S.C. was able to test his various engines, and try out new ideas in locomotive construction. There is no doubt that he was one of the great pioneers of the model steam locomotive world, and as a result of his articles and locomotive building, the model steam locomotive became much the most popular of all engineering models. Even today (and this is written in 1977), his designs are extremely popular, and locomotives built to his drawings are working in many countries throughout the world.

Another great pioneer in the model engineering world was Edgar T. Westbury. Roughly contemporary with L.B.S.C. Westbury was born in 1896. He served for a short time in the Royal Navy, being present at the

Battle of Jutland. Westbury designed and built his first small petrol engine while still serving in the Navy. In the late 1920s he was an Instructor at R.A.F. Cranwell and later at R.A.F. Halton he produced his famous 15 c.c. petrol engine *Atom Minor*, and in collaboration with Colonel (then Captain) C. E. Bowden, flew a model biplane with great success.

Other designs by Westbury came thick and fast, including the outstanding 5 c.c. *Kestrel*, which was the first small petrol engine to be fitted with a rotary induction valve. He worked alongside Sir Frank Whittle during the latter's early experiments with the jet engine; he was also a friend of T. E. Shaw, better known as Lawrence of Arabia.

During the Second World War, Edgar Westbury developed a number of small petrol-driven generators for use by the armed forces; he also assisted in the design of the famous Vincent-HRD motor-cycle. But Westbury's knowledge and skill was not confined to petrol engines. He produced a large number of most successful designs for small stationary and marine steam-engines and also items of workshop equipment, such as his well-known ¼ in. capacity sensitive drilling machine

The late George Gentry, a former member of the "Model Engineer" staff and regular contributor for many years.

Another shot of the Duke of Edinburgh during his visit to the *Model Engineer* Exhibition inspecting one of the trade stands. With him is the late Kenneth Garcke, then Chairman of Electrical Press, and a director of B.E.T. Ltd. On the right is J.N. Maskelyne.

and his light vertical milling machine, a modified version of which, known as the Dore-Westbury milling machine, has been built in many countries all over the world. A more recent design was for a three-cylinder high-speed radial engine for use in steam cars, using flash steam at high temperatures and pressures.

Westbury's last two designs were for a model steam fire-engine and for a Robinson type hot-air engine; the latter unfortunately was not a complete success, but even the most outstanding designer must be allowed one failure!

Edgar Westbury was a member of the (then) Junior Institute of Engineers and the Society of Model and Experimental Engineers. He was for many years President of the Model Power Boat Association and a Vice-President of the Sutton Model Engineering Society, the Romford Model Engineering Club and many others. His prolific writings on an enormous variety of subjects must surely remain as a source of sound technical information and indeed inspiration for many future generations of model engineers. Westbury never allowed himself much relaxation after he had retired from his position as technical editor to the *Model Engineer*, but was working right up to the end of his life, on new articles and new designs.

I first met Edgar Westbury in 1954, shortly before I joined the staff of *Model Engineer*, and always found him a constant source of help and information. At work, he was always good tempered, and his keen sense of humour helped all the staff over many a difficult hurdle. In fact I can truthfully say that in the whole sixteen years that I knew him, we never had a cross word. Perhaps Westbury's greatest quality was his ability to see the good points in other people's designs, and to give praise where due. He was indeed a great model engineer and a fine character.

A model engineer of a very different stamp was John Neville Maskelyne, or 'J.N.M.' as we used to call him. Actually, it is not really correct to describe J.N.M. as a model engineer at all, although he was intensely interested in model engineering and was in fact at one time a technical editor to *Model Engineer*. Perhaps a better description of J.N.M. was as a locomotive enthusiast, for his interest in and knowledge of the railway steam locomotive was quite astonishing.

J. N. Maskelyne was born in the last year of Brunel's 'Broad Gauge' and shortly before his death he wrote an article on the *Evening Star*, the last steam locomotive to be built for British Railways; it was published in *Model Railway News*. He was educated at a private school in Wandsworth from where he went to St. Paul's School, and then to King's College of London University to study engineering. Although, as he claimed, Maskelyne was a railway enthusiast from the age of one, he did not go to work on the railway, but received technical training with Vickers at Erith in Kent, and later joined the firm of Waygood-Otis, manufacturers of lifts and escalators.

Maskelyne's grandfather invented the differential-spacing typewriter, he was famous as an illusionist, as was his son Jasper. At the time of Jasper's death, J.N.M. had been at Waygood-Otis for twelve years. When J.N.M. left in 1930 to become a private consultant, he was already a regular contributor to *Model Railway News*, and in his *Real Railway Topics* revealed both his great love and knowledge of railways, and especially the steam locomotive. In January 1936, he accepted the editorship of *M.R.N.*

Under Maskelyne's chairmanship, representatives of all the interests concerned, meeting in London during the last war, formed the British Railway Modelling Standards Bureau to lay down suitable standards for the smaller gauges and scales of model railways, as had been done in the U.S.A. some years before.

In addition to his contributions to *Model Railway News*, J. N. Maskelyne acted as a consulting technical editor to *Model Engineer* for some years; he retired in 1957. Today, he is best remembered for his beautiful locomotive drawings, many of which were published in *Model Engineer* and in book form under the title *Locomotives I have known* and later *A further selection of Locomotives I have known.*

The late Gems Suzor, a well-known French model engineer, noted for his success with miniature petrol engines and hydroplanes.

I first met J. N. Maskelyne when I joined the firm of Percival Marshall in 1956 as a rather junior editorial assistant. My very first task at Noel Street, off Oxford Street, London, where Marshalls then had their offices and workshops, was to build a gauge '1' steam locomotive to an early design by Maskelyne, a simple free-lance 4-4-0 locomotive, with outside cylinders, inside slip-eccentric valve gear and methylated-spirit-firing. I am glad to be able to say that the engine was a success, being capable of hauling at least a dozen bogie coaches with no difficulty at all. It was intended for a 'scenic' type of model railway, and had no pretensions as a live passenger hauler, though it was criticized in some quarters as being old-fashioned — which of course it was!

After I have completed *Newbury*, as Maskelyne's 4-4-0 was duly named (he was then living near that Berkshire town), I designed and built the 3½ in. gauge L.M.S. Stanier 2-6-4 tank locomotive which was later to become well known under the name *Jubilee*, and I had much good advice and encouragement from Maskelyne while my drawings of this engine were on my board.

Maskelyne, like Percival Marshall and Henry Greenly before him, was a great personality of the modelling world, a tall, spare, old-world figure, always immaculately dressed in dark suit and wing collar, typical of Edwardian grace and good manners — I never heard him utter a harsh word. I can see him now, with neatly rolled umbrella and dark Homburg hat, leaving the office for Paddington Station, for the railway he loved above all others, and for his country cottage at Midgham in Berkshire. His like will not be seen again.

My last personality is none other than K. N. Harris, universally known as 'K.N.', now alas no longer with us. Apparently, he was known as 'Bucky' to his non-model engineering friends, and it must surely be unusual for someone to have two quite different nicknames!

K.N. was educated at a Leicestershire grammar school, where, in his own words, he entered at the bottom of the list but left at the top, mainly due to the influence of two remarkable teachers, who gave him his life-long love of the English language and his absorbing interest in all things mechanical. This probably accounts for the astonishing amount of writing he did, mainly for the *Model Engineer*, over a period of no less than sixty years.

Harris served an engineering apprenticeship with a Leicestershire firm of general engineers, which possessed a pattern shop, a foundry, and a constructional steel workshop, as well as a machine shop for the manufacture of boot and shoe machinery, also steam-engines and general engineering work for individual customers, so that he obtained a thorough grounding in basic mechanical engineering. Soon after his apprenticeship had ended, K.N. spent six months at sea as a junior engineer, to widen his experience.

Apart from model engineering, K.N. was a keen follower of rugby football, and a constant spectator at Twickenham; in his younger days he had been a keen rugby player and also a rowing man of some distinction. Later in his career, he joined the staff of Kodak Ltd., at Harrow, and it was while there that he formed the Kodak Society of Experimental Engineers and Craftsmen, of which he was the first Chairman and later Vice-President.

K.N.'s model engineering interests were many and varied and he won trophies and medals for his stationary and marines engines at many Model Engineer Exhibitions, though I do not think he ever completed a model locomotive. His last major exhibit at the Model Engineer Exhibition was a launch engine with reversing gear and feed pump, based on a design for his employers some fifty years earlier. This model was awarded a silver medal and the Bradbury-Winter Memorial Trophy at the 1959 Exhibition.

K.N. was a big man, in all senses of the word; though he had a rather gruff manner, this really concealed a heart of gold. He was very good company and was always very outspoken whatever the topic of conversation. Whatever subject was brought up, he always seemed to have made up his mind upon it. In his writing, he was very critical of L.B.S.C., for which he in turn, was criticized at the time. The casual observer might have thought that K.N. had a real personal dislike for 'LBSC' but I don't think this was really so. I think the true position was that 'K.N.' thought that LBSC's regular articles on building locomo-

The late J.N. Maskelyne's home, near Midgham, Newbury, Berks.

tives made the subject appear far too easy, which he thought could be very misleading. K.N. himself was a perfectionist where model engineering was concerned, while LBSC's approach was 'it works, so why worry too much about the methods used to make it?'

K.N. was sometimes criticized for apparently trying to introduce politics into our hobby. His critics claimed that politics had no influence at all on model engineering, and that the subject should be avoided at all costs. But K.N. insisted that politics had an enormous effect on our hobby (and in fact on all hobbies) and as just one example, quoted the effect of Purchase Tax or V.A.T. on the cost of our models, materials and so forth. I am afraid he was right!

Two little incidents concerning K.N. may give the reader some insight into his character. In conversation with him some years ago, I happened to mention that his old 'opponent', LBSC, had been taken ill, and had to temporarily discontinue his locomotive articles in *Model Engineer*. K.N.'s instant response was — 'Hope it's serious!' Of course, K.N. didn't mean this; in fact I think he would have missed his regular verbal 'battles' with LBSC. It was just an example of his caustic sense of humour. I'm glad to say that LBSC's indisposition at this time proved only short lived and that he was soon back in his 'locomotive works'.

The other incident was when he had invited me to have lunch with him, down in Sussex, somewhere near his home in Rustington. K.N. said that we must have some wine with our meal. I, somewhat tentatively, suggested a half bottle of something or other. K.N. looked quite hurt. 'A *half* bottle!' he cried, 'What's the good of a half bottle?' and promptly ordered a whole bottle.

"I sure love this hobby!"

Americana and The Iron Horse

Typical of the "Iron Horse" tradition is this fine model complete with "cowcatcher", bell and in the bright colours of the early period when the West was won.

Scene at Lord Gretton's Stapleford Park Railway, where his American locomotive "Lady Margaret" is getting up steam preparatory to the author taking the train.

A ¾ in. scale model of the C.N.R. 0-8-0 "Switcher".

Front view of Chet Peterson's Union Pacific 4-8-4 Northern, complete with brass bell and smokebox door headlight.

A 3 in. scale Case traction engine built by E. Ohlenkamp of Chicago, U.S.A. It is fitted with an all-copper boiler and weighs 700 lb. in working order.

An early American type 4-4-0 locomotive. The 19th century American engine was generally highly decorated, contrasting with the sombre plain black of later years.

AT THE BREAKER'S YARD

A rusting pile of icy steel
A pool of grease, a broken wheel
Your fire is dead, your boiler cold
But once so proud, so strong and bold
Your task to man is served and done
God bless my friend, from everyone.

(Terry Bird)

The Unusual

A model of a deep sea diver's helmet made by Mr. W.C. Holbird of Grays, Essex.

Another unusual model — a Ransomes thresher made by Mr. A. Buckley of Dewsbury.

An unusual model — a Tram engine based on a Burrell prototype, made by T. Morris.

An overhead view of a fine model steam fire engine built by Miss Cherry Hinds, of Malvern, Worcs.

A model 4-cylinder side-valve petrol engine to the Westbury "Seal" design.

Ann Carter (now Mrs. Hatherill) made this spinning wheel.

A remarkable ½ inch scale model of an oscillating cylinder paddle engine by the late Cdr. W.T. Barker.